Smart Business Solutions

Direct Marketing and
Customer Management

DOUGLAS GANTENBEIN

PUBLISHED BY
Microsoft Press
A Division of Microsoft Corporation
One Microsoft Way
Redmond, Washington 98052-6399

Library of Congress Cataloging-in-Publication Data
Gantenbein, Doug.
    Smart Business Solutions for Direct Marketing and Customer
    Management : How to Harness the Power of Technology to Put Your
    Small Business in the Black / Doug Gantenbein.
        p. cm.
    Includes index.
    ISBN 0-7356-0683-8
    1. Small business--Technological innovations. 2. Direct
marketing--Management. 3. Small business--Management. 4. Consumer
satisfaction. I. Title.
HD2341.G26   1999
658.8--dc21   99-14925
                                            CIP

Printed and bound in the United States of America.

1 2 3 4 5 6 7 8 9 MLML 4 3 2 1 0 9

Distributed in Canada by Penguin Books Canada Limited.

A CIP catalogue record for this book is available from the British Library.

Microsoft Press books are available through booksellers and distributors worldwide. For further information about international editions, contact your local Microsoft Corporation office or contact Microsoft Press International directly at fax (425) 936-7329. Visit our Web site at mspress.microsoft.com.

Reproduction of Model 88 self-mailer used by permission of Cambridge Soundworks. Synopsis of Power*Test* used by permission of Robert C. Hacker, President, The Hacker Group Ltd. Reproduction of catalog copy used by permission of L.L. Bean, Inc. Reproduction of direct-mail postcard used by permission of Paraform, Inc. Reproduction of direct-mail envelope used by permission of Schwartz Brothers Restaurants.

FrontPage, Microsoft, Microsoft Press, Outlook, PhotoDraw, PivotChart, and Windows are either registered trademarks or trademarks of Microsoft Corporation in the United States and/or other countries. PANTONE is a registered trademark of Pantone, Inc. Other product and company names mentioned herein may be the trademarks of their respective owners.

The example companies, organizations, products, people, and events depicted herein are fictitious. No association with any real company, organization, product, person, or event is intended or should be inferred.

For Microsoft Press
Acquisitions Editor: Christey Bahn
Project Editors: Kim Fryer; Jenny Moss Benson

For Benchmark Productions, Inc.
Project Manager: Amy Yeager
Production Manager: Greg deZarn O'Hare

Contents at a Glance

Contents

Part 2

Customer Management That Builds Sales and Loyalty

Part 3
Appendixes

Acknowledgments

Many people helped with valuable insight and assistance during the production of this book. While I'm surely overlooking some names, I'd like to acknowledge the help of direct-mail experts Jackie Bailey, Carol Worthington, Robert Hacker, Bob Morris, Debra Jason, and Carol Coture. In the section on customer service and client management, I'm particularly grateful for the input and assistance of Jay Goltz, Lee Duffey, Kathy Leck, Mary Jane Pioli, Marilyn Hawkins, and Elizabeth Spaulding.

Douglas Gantenbein
April 23, 1999

Part 1

Direct Marketing That Gets Results

Chapter 1

Direct Mail and the Small Business

Sur la Table is a Seattle-based store that is a cook's paradise. Whether it's the perfect potato peeler, the ideal roasting pan, or the latest Italian coffee-making accessory, a customer can find it at Sur la Table. From its start 27 years ago in Seattle's famed Pike Place Market, the kitchen-supply store has grown into a seven-store chain, with outlets in the Seattle suburb of Kirkland, Newport Beach, San Francisco, Berkeley, and Santa Monica in California, and now Dallas, Texas. Sur la Table faces a tough marketing environment as it competes against nationally known names such as Williams-Sonoma as well as regional cooking stores. And getting the word out is difficult. "We're a small company," says Carol Couture, Sur la Table's director of marketing. "We're not big enough to afford broadcast every day or do run-of-press in a newspaper and hope that someone opens the food section."

The marketing solution for Sur la Table: direct mail. Each year the company sends out 7 million catalogs as well as local letters to customers about in-store cooking classes and newsletters to tell customers about upcoming special events and demonstrations. The catalogs generate nearly half the company's revenue, create a large portion of Sur la Table's in-store traffic, and help the company target the best retail locations by providing feedback on customer location and demographics.

"In my mind, direct mail is the most efficient way to grow a business," says Couture. "I love being able to send a piece to someone's house and to predict what the response might be."

Sur la Table is but one of thousands of small businesses in the United States that have found direct marketing to be the most efficient and effective way to spend their marketing dollars. Chances are, it's the right approach for your business too, whether you own a corner floral shop, a chain of drive-through espresso bars, or a local clothing store.

What Direct Mail Can Do for You

Advertising is expensive. Take, for example, the 1999 Super Bowl. Advertisers spent as much as $1.3 million for as little as 30 seconds of air time. For that astonishing sum, advertisers were rewarded with a huge audience—128 million people tuned in to see whether the Denver Broncos or Atlanta Falcons would prevail. The ads that ran during the Super Bowl, though, were the marketing equivalent of a shotgun—broadly aimed, perhaps even barely aimed. They were solely the province of advertising's 500-pound gorillas, companies such as Budweiser, Federal Express, and M&M/Mars. No Sur la Tables there, nor any other small businesses.

For the small business, direct mail is the best way, as Carol Couture says, to "swim with the big fish." And to do so without their advertising millions. Here's why.

Direct Mail Creates a Response

Although it can do several things, such as generate store traffic or create leads that might later become sales, direct mail primarily is aimed at getting a *response*. No other form of advertising works as well at coaxing the consumer to take action. If you want a customer to visit your restaurant, purchase a sweater, try a carpet-cleaning service, or subscribe to a magazine, direct marketing is the most effective way to achieve your goal. That's a particularly important point for a small business, which does not have the luxury of waiting for an advertising campaign—even if it could afford one—to work its way slowly into the customers' consciousness.

Direct Mail Grows a Customer Base

In 1967, L.L. Bean had annual sales of $3 million. Twenty-four years later, in 1991, it hit the $1 billion mark. How? Largely by taking advantage of powerful new database tools that allowed Bean to more carefully target and accurately measure its catalog sales campaigns. L.L. Bean is just one of hundreds of companies that have used direct marketing to greatly expand a customer base. Others include Lillian Vernon, Banana Republic, and Sharper Image—not to mention Sears & Roebuck, one of the first companies to understand the power of direct marketing.

Direct Mail Can Be Tailored to Your Business

As a small business, you have a particular set of goals and ideals. That you're an entrepreneur also means you have a particular business personality. Direct marketing can easily be designed to fit your business just as stripes fit a tiger. You can be aggressive, or folksy, or professionally slick, or casual and informal.

Direct Mail Can Be Measured

With most advertising, it's difficult to know if the ads have any real benefit. Has the public's perception of a company changed? Are customers more aware of a brand? Direct mail, on the other hand, can be precisely measured. You can easily calculate the cost of a mailing and the cost per response, and you can weigh those costs against the sales generated by a direct-mail campaign.

Direct Mail Can Be Tested

What offer works best? What design works best? Are flyers better than catalogs? With direct mail you can find out. Big catalog retailers such as Lands' End and L.L. Bean, for instance, send catalogs to the East and West Coasts with different offers—a different price, or a product that is given special emphasis, or even one cover for the woman in the household (with more colors) and another for the man (more businesslike). The sales figures then tell the story of whether that special price or special catalog placement was effective. (As a consumer, you *always* want to ask if a catalog item is on sale; it very well might be in a different catalog from the same company.) A small business can test many variables, too. It can change the copy in an ad, experiment with different formats, or change the offer from a 10 percent discount on an oil change to a free oil change with a complete service checkup.

Direct Mail Can Fit Any Budget

Direct mail can be tailored to fit the size of your budget, whether you want to mail postcards to 500 customers, reminding them to rotate their tires or change their antifreeze, or print and distribute a full-color catalog of your new clothing designs. It's easy to scale up a direct-mail campaign, so if results from the first 10,000 mailings are good, ordering another 50,000 copies, purchasing or renting some expanded mailing lists, and making the campaign even bigger are simple.

Direct Mail Is Personal

A properly executed direct-mail campaign reaches a carefully chosen set of customers. Because of that, you can be very specific when you discuss how the product or service will address the recipient's needs. Your mailing piece and its copy also can be very direct and personal in addressing the customer, unlike a mass television, radio, or print campaign. This personal approach leads to one of the real strengths of direct marketing: its ability to help you develop a strong, long-term relationship with a customer. By tracking your customers' purchase histories, you can offer special incentives to your best customers, remind lapsed customers that you'd still like their business, or tailor offers and products to suit particular customers.

Direct Mail Is Hot

In 1998, according to the Direct Marketing Association, direct-marketing sales (which include direct mail and telemarketing) to consumers totaled $759 billion, or 12.4 percent of total U.S consumer sales of a little more than $6 trillion. (No wonder they say ours is a consumer-driven economy.) Direct marketing also scores well in the business-to-business category, with $612 billion in 1998 sales, 5 percent of the national business-to-business total of $12 trillion.

Direct-marketing sales are growing, too, by an average of nearly 8 percent a year since 1993. Although that growth rate has been tracked fairly closely by cost increases, direct-marketing experts predict that cost growth will decrease in the next few years even as sales growth increases to about 8.6 percent annually, resulting in strong earnings for direct marketers. In all cases—advertising spending, revenue, and employment—growth of direct marketing is expected to outpace U.S. economic growth.

Direct marketing accounts for a huge percentage of revenue for those companies that engage in it. Another recent survey, conducted by *Direct* magazine, showed that companies using direct marketing earn nearly half their revenue from that approach. In the case of business-to-business marketing, the figure is more than half, closely followed by consumer marketing. Twenty-three percent of the survey's respondents reported that direct marketing was responsible for *90 percent* of their total sales.

Direct Marketing: Fact and Myth

Statistics such as those just discussed debunk one common perception of direct marketing: that it doesn't work. It's true that some consumer surveys may indicate annoyance with the sheer volume of direct-marketing pitches many people face. Other surveys show that in any given year more than half of all Americans purchase something by phone, by mail, or via the Internet. Those are all direct-marketing sales, folks. It does work—and it works well. Consumers who have clipped coupons, ordered from catalogs, kept and used discount cards mailed to them by a restaurant, or checked sales at local stores that sent them postcards have responded to direct marketing.

Lois K. Geller, a New York–based expert on direct mail and author of the book *Response: The Complete Guide to Profitable Direct Mail*, lists some other common misconceptions about direct marketing:

It's easier to sell retail. Not true, says Geller. Even though the retail setting seems to give you better access to the customer, you may actually have little control over where the product is placed in a store or how a store's sales staff presents it to a potential buyer. Moreover, selling retail forces you to manufacture enough of an item to fill shelves at multiple outlets. In contrast, direct marketing gives the seller great control over how the product is presented. It can also be much more efficient, allowing the seller to manufacture or purchase more of an item only as needed. This "just-in-time" production approach is what helped Toyota to succeed and what led Ford and other car makers to follow suit. You can benefit from the same approach when selling to a few thousand people.

I already advertise, and I'm doing well. I don't need direct mail. Well, maybe you don't, says Geller. But how do you know your advertising is what's drawing customers? Send out a coupon that offers a 25 percent discount, and you'll know exactly what sort of response your advertising dollar—when used in direct marketing—is earning for you. Otherwise, you could be wasting advertising dollars if location and word-of-mouth alone could generate traffic.

I tried direct marketing, and it didn't work. Fair enough, says Geller. But are you sure you got your pitch into the right hands? That you had the right offer and the right product to sell via direct marketing? Sometimes a detail as minor as the wording of a headline can make a difference. Try several approaches; one of them is almost certain to bring results.

Great Moments in Direct Mail

1450: Gutenberg invents movable type.

1667: English gardener William Lucas publishes earliest gardening catalog.

1744: Benjamin Franklin publishes a catalog "near six hundred volumes in most faculties and sciences." Franklin further develops the concept of guaranteeing customer satisfaction.

1830s: Several New England companies begin to sell sporting equipment, fishing gear, and marine supplies.

1867: Invention of the typewriter makes it possible to print small quantities of material cheaply and relatively quickly. Alas, early models struck paper from the bottom, making it impossible to proofread.

1886: Richard Sears, a railroad station agent, enters the mail-order business to sell watches refused by an addressee. He joins with Alvah Roebuck in 1887 and six years later prints a 196-page catalog.

1912: L.L. Bean founds a successful mail-order company on the promise of a guaranteed waterproof boot.

1926: Neiman Marcus mails the first catalog with expensive, high-end clothing and gifts.

1950: Diners Club mails out the first credit card.

1982: Dr. Roger Breslow, a New York internist, keeps every piece of direct mail he receives in a year. His collection weighs 509 pounds.

1992: The number of Americans shopping from home exceeds 100 million.

OK, It Isn't All Great News

Direct marketing is a proven, effective, and growing method of spending your marketing dollars. But all is not instant profits and cheer. There's the "J" word, for instance. That's right, junk mail. Even people in the industry use it—at least, most of them do. Bob Hacker, a Seattle-area direct-mail consultant whose campaigns for clients such as IBM have resulted in tenfold response rate increases over previous efforts, notes one unavoidable fact: "Nobody," he says, "likes direct mail." In fact, in even the most successful campaigns 90 percent of the people you reach *will throw your message away.* That's right. Throw it away. As garbage. In a reasonably successful campaign that figure jumps to 98 percent.

One reason might be that people sometimes seem overwhelmed by direct mail. In 1998, for instance, catalog mailings jumped by 15 percent, according to the *Wall Street Journal*, with the typical "prime" catalog customer—a working couple with young children—receiving nine catalogs a week. Many stores that were strictly bricks-and-mortar retailers only a few years ago, such as Saks Fifth Avenue and Dillards Department Stores, now have entered the direct-mail fray with catalogs. Increased competition has made it more difficult for even well-established direct-mail marketers to maintain growth and profit rates. The brutal competition of recent years savaged highly respected Lands' End, the catalog clothier, resulting in shakeups at the top and layoffs among the employees. Then there is the proliferation of credit-card offers, magazine subscription offers, and record club enticements. The fact is that it's difficult to cut through the clutter.

You should consider other issues as well before launching a direct-mail campaign.

Higher Postage Rates

Through the early 1990s, rates for third-class mail (usually used for direct mail) increased by 16 percent. Rates stabilized during 1997 and 1998, but they increased again early in 1999. Higher mailing costs, of course, increase a direct marketer's cost-per-thousand, meaning that each mailing must show an improved return to make a profit. What's more, an increasing number of direct-mail consultants say that using bulk mail all but guarantees that your piece will be thrown out. Some even use high-end carriers such as Federal Express to ensure that a mailing gets to a selected group of readers.

Higher Production Costs

Along with higher postal rates, costs for paper, printing, and other production-related items have gone up as well. The good news, as the Direct Marketing Association's 1998 industry report notes, is that those costs are expected to grow at a slower rate than sales during the next several years. Still, postal costs and production costs can account for up to 90 percent of the total cost of some mailings. Any jump here can be substantial.

Privacy Concerns

Face it—some people simply don't want unsolicited advertisements. To date the direct marketing industry has done a poor job dealing with this issue. It's true; consumers can have their names taken off mailing lists.

The Industry Itself

Sometimes direct mail shows a remarkable ability to take direct aim only at its foot. How many pieces of mail do people get each week that are stamped "Important," have fake postal-handling instructions all over them, and contain assumptions about the recipient that reveal the hand of a giant mailing list that does little to differentiate among potential customers' actual needs or wants? As *Direct* magazine columnist Herschell Gordon Lewis recently noted, too many direct-mail offers start out with patently absurd lines such as "Make $1,000 weekly stuffing envelopes!" or "New money-making concept makes all others obsolete!" or "I made over $1 million cleaning dirty mini-blinds!"

Wrote Lewis: "We're regarded as sharpies, as fast-buck artists. The great unwashed public, with its great unwashed mini-blinds, lumps us all into one cauldron. Direct marketing, direct response, direct mail, mail order, telemarketing—it's like that line from the old comic strip 'Pogo,' 'We have met the enemy and it is us.'"

The message is not that direct mail doesn't work. The message is that direct mail must be done well. And when it is done well, direct mail becomes a genuine customer service, not an intrusion.

Is Direct Mail Right for Your Business?

Direct mail's adaptability makes it suitable for nearly any business. Here are some examples of direct-mail campaigns from companies big and small:

- To help introduce its new M-class sport utility vehicle in 1997, Mercedes-Benz developed an elaborate direct-mail campaign, beginning in October 1995 with a 400,000 piece mailing sent to prospective customers (mostly well-to-do suburbanites). Using surveys and including in its mailings some interest-piquing items such as bits of sheet metal said to be from an M-class prototype, Mercedes-Benz generated response rates as high as 50 percent and created considerable buzz for the new $45,000 vehicles, which proved to be a smash hit for Mercedes-Benz.

- In Seattle, the locally owned Schwartz Brothers Restaurants perked up sales during the slow January-February period with a colorful mailer with a handy business-card-sized tear-off stub good for up to $66 in discounts at the chain's eight restaurants and delicatessens. The result was a 6 percent response, adding tens of thousands of dollars in revenue to an otherwise dead time in the restaurant business.

- In Westport, Connecticut, upscale clothier Mitchell's assembled a mailing list of customers who had not been in the store for two years and who tended to spend less than $900 for a suit. The mailing offered special discounts. Mitchell's sent out 3,000 pieces, generating 438 trips into the store—an impressive response rate of nearly 15 percent. Those returning customers spent more than $300,000, income for Mitchell's that far exceeded the cost of the mailer.

In short, direct mail can work for virtually any business, with offers and products big and small. Here are just a few of the things you can do with direct mail:

- Sell a product or service
- Generate store traffic
- Thank customers for business
- Tell customers about new merchandise, a new location, or new employees
- Sell supplies and accessories to recent customers
- Offer special discounts
- Conduct customer surveys
- Retrieve customers who have not shopped with you for a while

- Persuade good customers to be better customers
- Introduce yourself to prospective customers
- Announce a store opening
- Build a mailing list of potential customers
- Excite customers about an upcoming product

How Microsoft Office 2000 Can Help

Microsoft Office 2000 can be a valuable tool in developing and executing a direct-mail strategy. It includes useful upgrades to many of the programs you already are familiar with and that can help you write and design direct-mail pieces, such as Microsoft Word and Microsoft Publisher. In addition, Office 2000 Small Business, Professional, and Premium versions include the Direct Mail Manager. This powerful Internet-based tool uses Microsoft wizard technology and the expertise of the U.S. Postal Service to make it easy for small businesses to develop precisely targeted and cost-effective mailings. The Direct Mail Manager includes the following:

- Access to targeted prospect lists from InfoUSA, a respected provider of mailing lists for direct-mail users. Office 2000 users can employ the Direct Mail Manager to access InfoUSA's Web site to scan available lists and costs, and then purchase and download names directly into a Microsoft Excel spreadsheet. Special discounts and offers apply to users of Office 2000.

- Tools that allow you to verify addresses against the U.S. Postal Service national database. The database will check addresses for completeness and fix incorrect addresses.

- Word or Publisher mail merge. This feature helps users send their documents or publications or hand off mailing tasks to an outside mailing service such as Pitney Bowes or Neopost.

Direct mail, Office 2000, and your small business—a profitable combination.

Chapter 2

Plan Your Way to Direct-Mail Success

Harvey Penick, the noted golf instructor, once wrote that even duffers should get in the habit of going through all the motions of making a perfect golf shot: judging the distance, selecting a club, carefully aiming, and taking a controlled swing. They may well hit one of their usual lame shots, Penick wrote. But then again, because of that preparation they may well hit the perfect shot. His admonition: Any plan is better than no plan.

Creating a successful direct-mail campaign is a lot like taking a golf shot. It's true that even a well-planned campaign may not go as hoped, but it's certain that an unplanned or poorly planned campaign will fail. Evaluating your product, taking a careful look at your potential customer base, and determining the best message will go a long way toward ensuring that your campaign works as smoothly as a Tiger Woods golf swing. Without solid planning, no amount of clever copywriting or snazzy graphics will salvage the effort.

First, a Reality Check

Direct mail is one of the most effective marketing tools available to a small business. But it doesn't work miracles. Jim Morris, a direct-marketing consultant based in Denver, says that one of his first tasks with clients is to explain to them exactly what they should expect from a direct-marketing campaign. That's because most novice marketers believe that their product or service will revolutionize the marketing world. "They're just too close to their product," says Morris. "They think the response rate will be 50 percent once everyone sees how exciting their new product is. They figure the money spout will just open up."

Alas, that may not always be the case. Industry studies, for instance, consistently show that a 2 percent response rate is considered normal. There are hints that figure is dropping, in fact, as volumes of direct mail from catalog companies, credit card issuers, and other expanding direct-mail users load mailboxes. Still, some marketers remain aggressively optimistic. Morris remembers well one particular client. "She thought *everybody* would respond," he says. "She really did!"

Direct marketing is a powerful tool. It doesn't work for every product or service, though, and it may take several tests and offer revisions before a marketer gets a response rate that is worthwhile.

What to Sell?

As noted in Chapter 1, the list of items that can be sold or promoted by direct mail is almost limitless. Still, determining the right product to market via direct mail takes some thought. Let's say, for instance, that you own a local garden center that stocks a wide range of plants and trees as well as gardeners' supplies such as tools, fertilizers, and planting pots. What is it you sell? You may have a range of the newest, most exotic hybrid plants and 12 different gardening shovels. You may also stock many tried-and-true standbys such as rhododendrons and round-blade shovels. Which do you emphasize? Then there's pricing. Will you try to compete with a nearby "home center" on price, or will you emphasize your unusual, high-quality (but expensive) merchandise? Then there are your customers—or the customers you'd like to attract. Are they avid gardeners who love to spend the weekend digging in dirt up to their elbows? Or casual gardeners who would rather spend most of their weekends playing tennis or going for a hike?

Another scenario: You run a small public relations company and would like to attract new clients. Do you want to persuade businesses that already work with an agency to move to yours, or do you want businesses that have never tried public relations to hire you? Do you present yourself as a small, friendly agency or as a firm that's big enough and capable enough to handle major corporate clients? And what services do you want to sell? Writing? High-tech marketing? Image building?

In short, it's not an easy process. But most direct-marketing experts say you can ensure that your product or service is suited to direct mail by checking whether it meets the following criteria:

It has a targeted audience. You never see direct mail from McDonald's for a reason: Everyone already eats its hamburgers (well, Burger King would argue that point). Direct mail works best when you can identify and target a distinct set of potential customers who, by virtue of income, past buying habits, hobbies, or career, are apt to be singularly interested in your product or service.

It's neither too expensive nor too cheap. Direct mail respondents are fairly price-sensitive—in fact, most are looking for a "deal" when they leaf through the direct mail that arrives at their home. Pricing is an important factor. Too expensive, and people won't be able to make an impulse decision to purchase the item. Too cheap, and you won't cover your costs. There are exceptions, of course, particularly for high-priced items (say, $500 and up). Several car makers—including Mercedes-Benz—have had success with carefully targeted direct-mail campaigns. Same for Suzuki motorcycles. The key for both companies was to fine-tune their mailing lists so that only top prospects were contacted. It may also be possible to overcome price resistance with information. Before people spend large sums of money, most want to research a product thoroughly. If you send a detailed direct-mail piece that anticipates and responds to a reader's likely questions (and objections), you may have success.

It is an item that people regularly use. Most people in direct marketing could take a tip from their dentist or their beagle's veterinarian. Chances are, these small-business owners regularly send out reminder cards about a teeth cleaning or vaccination. In fact, just about any item or service a person is apt to need or use up on a regular basis lends itself to direct-mail reminder cards—particularly those that offer something extra, such as a discounted service or a two-for-one deal.

Writing Your Game Plan

After settling on the product you want to sell, it's time for some serious pencil sharpening. Take out a legal-sized sheet of paper and start writing down the answers to the following questions.

What are your product's strengths? Is it an improvement over a previous model? Does it offer longer wear? Is it smaller and lighter? Will it make the user wildly popular? If so, how? Spend some time really thinking about what your product can do. Try to see these advantages through a potential customer's eyes. Remember, your own eyes may be so infatuated with your product that you can't get beyond the words "It's perfect!"

What are your product's weaknesses? Of course, we know nothing is perfect. This includes your product or service. What are its flaws? Is it a little expensive? Cheaper, but not as fully featured as a competitor's product? It is time to be honest. It's better to admit to some flaws now so you can figure out how to defuse problems in your direct-mail piece, or even work now to fix those problems.

What are your competitors' strengths and weaknesses? Go through the same drill for the other guys. How does your product stack up against theirs? Read their literature carefully to see what claims they are making; you may get some insight into how to promote your own product advantageously.

How big is the potential market? Are there statistics that show how many people now use a similar product or might be interested in purchasing such a product? How many people in that potential market pool can you contact? Assuming a response rate of 2 percent, what number does that yield?

What do your customers look like? Who are they? How old, young, rich, poor? What do they do in the world? What motivates them? What problems do they face? What would drive them to buy your product? Fear? Greed? Guilt? Anger? A desire to stand out? A desire to fit in? Even if you're working on a business-to-business product, remember that it's a human being—not a corporation—who ultimately will make a decision about your product. And all human beings are essentially the same.

What is your objective? Sure, it seems like a simple question. But really, what do you want to do? Bring in sales? Generate foot traffic? Get people to think about your product the next time they see it in a store? Qualify possible contacts for your sales force? The answer to that question will determine much of the content of your actual direct-mail piece.

What is your budget? How much can you afford to put into a direct-mail campaign? The answer to that will depend on several factors, including the cost of the item you're selling, the per-piece cost of the mailing, and the expected response rate. Don't assume that you'll sell 10,000 units of a product at a profit of $10 each and then base your direct-mail budget on that number. It is better to run a small, carefully controlled test to get a more accurate estimate of how people will respond. Then you can scale up your mailing to accommodate the anticipated return on your investment.

Are you prepared for the response? Of course, you don't want to overestimate the potential response to your mailing. Nor do you want to underestimate it. Are you prepared to produce 2000 or 5000 or 10,000 units of your product if the response is strong? And to do so quickly? Customers today are accustomed to having products in their hands quickly. Shoppers at big catalog merchandisers such as L.L. Bean and Lands' End, for instance, expect to have their purchases shipped to them within a few days. If you make customers wait as little as a week to receive your product, you run the risk of alienating them or even losing the sales should they get so exasperated that they refuse to accept delivery.

What next? Here's an age-old piece of marketing wisdom for you: It's easier to keep old customers than to attract new ones. Obvious as a peacock in a chicken house, right? But many direct-mail novices get so wrapped up in the problem of the day (production, mailer development, postage hassles) that they don't think ahead to a point a year or two away. Direct-mail veterans recommend that a marketer be prepared to capture the names and addresses of everyone who responds to a mailing piece and to have in place a continuity program such as a follow-up mailer, a method to approach a buyer about upgrades or service, or an extended warranty. You don't want people to make one purchase and then drop off your radar screen. You want them in your sight for years to come.

Writing the Offer

One thing separates direct-mail marketing from all other forms of advertising: its dependency on what's called the *offer*. Budweiser sells beer with talking frogs and lizards; you don't have the budget or time for that sort of immensely expensive, image-building kind of advertising. You need to *sell* your product, and the best way to do that—once you've targeted your potential market—is with a strong offer. Potential customers may find your product appealing, your new restaurant convenient, your service useful—but it is the offer that will get them off the couch and either writing you a check or reaching for a credit card. That is, after all, exactly what you want them to do, right?

Make no mistake: Your direct mail piece *must* have an offer. A customer may wander into a new restaurant or take a look at your product if he or she happens to come across it, but a good offer will vastly improve the chances that he or she will visit that restaurant or purchase that product. Says direct-market sage Mary Kerford, a vice president with Cohn & Wells in San Francisco: "You want to use that offer to really generate the feeling of 'I've gotta have that *now!*'"

The Four Kinds of Offers

For starters, says Kerford, you want to understand the four different offer types. Again, look through that box of direct mail you've collected in the past week. You're sure to see variations on these offers throughout that pile.

Value-added. Automobile makers such as Toyota, Ford, and Dodge all use this type of offer, and so can you. Simply put, when customers buy your product, they get something else. In the case of an automobile, the offer may be a free accessory or set of accessories—fog lights, for instance, or a hood bra. If it's an item that might be useful to a customer in multiples (one for the customer, one for a spouse) the offer might allow purchase of a second item at half price or a third item for free if two are purchased.

Price-off. One of the most common offers, the price-off variant is a straight discount: "Order now and save 20 percent!" Not rocket science, but it is consistently successful at selling things. All people, deep down, want to think they got a deal.

Free trial. People often hesitate to purchase an item through a mailing because they can't wrap their fingers around it, heft it, or show it to a friend without actually buying it. Encourage hesitant buyers by assuring them that they can try a product for 30 days or 60 days or 90 days absolutely free. That gives the buyer an "out" should the product not prove satisfactory. In reality, studies show that very few purchasers return a product once they have it in hand, opting to buy it instead. This technique has proven particularly popular with software makers, who offer free downloads of their software. The software contains a built-in "timer"; at the end of a specified period it disables the software and asks the purchaser to call the company. In exchange for a purchase, the buyer receives a code that enables him or her to keep using the software.

Additional information. Sometimes the initial offer is simply meant to further qualify the prospect. Particularly for big-ticket items or other major financial decisions that require large, complex, expensive mailing pieces, it's worthwhile to begin the transaction by piquing the prospect's interest and asking him or her to respond in return for additional information. After all, if someone takes the time to *ask* for more direct mail, it's fairly certain that he or she is genuinely interested in the product or service. Stockbrokers and high-tech companies are among those who take advantage of this technique. It can also be used to generate traffic at a trade show; for instance, it might promise that a company salesperson will review a product line with any interested customer who stops by the booth.

Strengthening the Offer

Within those four basic offer types, there's plenty of room for putting your own spin on an offer, or for mixing and matching any of the types. Most of all, though, you want to ensure that your offer is as strong and compelling as you can possibly make it. Here's how to take any of the four basic offers and make it one a customer can't—well, is hard-pressed to—resist.

Make it good

It's true that when you're offering prospective customers a discount or a second item for free or half price, you're taking money out of your pocket and putting it in theirs. Any small-business owner will tell you what a painful operation such a cash transplant can be. You have to endure a little pain, though, to make direct mail work. In other words, make the offer a good one. Ten percent off is old hat; 25 percent off is apt to get someone's

19

attention. And don't forget: "Free" remains one of the most powerful words in the English language. If you can make any part of your offer free, the offer will be better for it.

Make it involving

You know how you sort through mail: standing over a trash can, making split-second decisions about whether to keep anything you don't immediately recognize. How do you keep a prospect from doing the same when the mail contains your marketing piece? By having a good offer and making it *involving*. Involvement means the reader thinks about it, does a little math to weigh the item's cost against his or her current checkbook balance, and then makes a little mental leap into the land where he or she owns the product.

How to make it involving? Anything you can do to keep that envelope or letter suspended over the round receptacle is to your advantage. One tip: If you're offering percent-off, make it a number that makes the reader stop and think for a few seconds, such as "34 percent off!" That's more attention getting than a round number, and potentially more involving. So too is an offer of one item with a second at half price, or two items with a third for free. That leads readers to start calculating how much they'll save if they buy multiple units of your gadget. When they do that, they're well on their way to actually making a purchase.

Make it simple

In addition to thinking carefully about the product or service, the prospective direct marketer must remember two things. One is that most direct mail pieces sell only one item (catalogs are the exception, of course). It may be a single selling proposition that covers an entire product line (25 percent off all inventory!), but you want to build your direct-mail piece around a single product or offer. The other key point is that simplicity works best in direct mail. You don't want to clutter a mailing piece with all sorts of options, trying to be all things to all customers. Build a single, compelling reason for a customer to purchase your products or services.

In fact, during the next week, set a basket or cardboard box next to the desk where you sort your mail. Into it, toss all the direct-mail pieces that come your way. At week's end sort out the ones that pique your interest or that made you stop to think about whether to purchase the item offered. Chances are you'll find that these pieces adhere to the two guidelines just discussed. Some examples follow:

- The 3M Company wanted to draw traffic to its booth at the winter Outdoor Retailer Show in Salt Lake City. In particular, the company wanted people to see its new stretchable insulation for use in clothing. 3M sent a postcard touting the insulation and their booth location to past attendees of the show.

- Magnolia Hi-Fi, a Seattle-based retailer of high-end audio and video systems, wanted to bring in store traffic and boost sales during the quiet post–Super Bowl period. It sent a postcard to customers on a list drawn from the company database that simply said, "Our best prices. Our best financing. For our best customers!" The word "best" was emphasized, making it the central theme in the card.

- Hyatt Resorts, in the hope of getting summer travelers "locked in" well ahead of its competitors, ran a campaign in early 1997 with the tag line "Ready, set, stay." Through eye-catching photography and a value-added offer (a free night's stay), Hyatt was able to generate a response rate of nearly 6 percent. Resort guests who were drawn to Hyatt by the campaign spent an average of $1,200 per visit.

Make it relevant

Timing is important in direct mail. If you plan a mailing in November or December, for instance, you'd better be prepared to compete for attention with the thousands of other mailings going out during those months of holiday shopping. Advertising back-to-school specials at your clothing store? Then send the mailing in early August, when parents are beginning to plan for the start of school in late August or September. And if you're sending a business-to-business piece, keep in mind that Tuesday and Wednesday usually are the lowest-volume mail days. To ensure that your piece arrives then, do your mail drop on a Friday.

Make it believable

Never forget this: People are remarkably clever. Sure, there's the guy who drives to the Publisher's Clearinghouse office for his $1 million check because, after all, the envelope *said* he was a winner. Count him as a minority. The vast majority of people who will read your direct-mail piece have heard it all before. They've been conned, overpromised, and given the Asterisk that Taketh Away so many times that if those stunts were nickels, they'd be retired in a villa in Tuscany. So, make your offer believable. If it's too good, people will figure there's a catch.

An Offer Isn't Always Just a Deal

The offer primarily is the best deal you can provide on your product or service. There are other sorts of "offers" you can make as well:

A time limit. A common trick, but a good one. "Urgent!" your envelope might say. "90 days to respond!" Or "Offer good this month only!" The key is to break through recipients' complacency. If they think they can stroll in just any old time and get a big discount, they might *never* stroll in.

Payment options. Pricing psychology can be vitally important. Just about any direct-mail expert will tell you: A $99 offer will beat a $101 offer almost every time. Sure, it's only two dollars, but it can make all the difference. Make the purchase as easy as possible by emphasizing the smallest figure possible. One way to do that is to allow a customer to make three $33 payments. Be careful here: You can't be "loaning" customers the money and asking them to pay you back or adding a little bit for interest. Do that, and you've become a credit company. There's no reason, though, not to make it a straight $99 offer, with three equal payments.

Premiums. It's common nowadays to order a magazine subscription and receive something extra—a logo T-shirt, a sports bottle, perhaps even a calculator or personal stereo. Depending on your product and profit margins, it might be worth your while to offer buyers a premium. This can be tricky, however. It can't be too expensive for you, but it must have *perceived* value for the customer. Then there's the question as to whether to offer a premium that complements your product (for example, a die-cast metal truck for youngsters whose parents have their oil changed in your shop) or one that is simply a popular item (a Beanie Baby). It's perhaps best to make the premium attractive, but fun. Give a premium that reflects your personality and the spirit of your company. And keep in mind that premiums can open pitfalls; the Federal Trade Commission has guidelines on just what constitutes "free." Check with the FTC to get more information.

Creating an Analysis Worksheet

Microsoft Office 2000 is designed to make it easier than ever before for the entrepreneur to apply the intelligence of Office to scores of common business problems. For planning a direct-mail campaign, Microsoft Excel is a particularly powerful tool. New features for Excel 2000 include the following:

- **List AutoFill.** Excel 2000 automatically extends formatting and formulas in lists, simplifying common tasks.

- **See-Through View.** The See-Through selection in Excel 2000 lightly shades cells so that you can see the results of your changes without unselecting cells.

- **Chart improvements.** Excel 2000 features improved formatting and the ability to create PivotChart reports for more dynamic views. With PivotChart reports, you can link sets of data more easily for sophisticated analysis.

Excel's ease of use and powerful analytical capabilities make it easy to evaluate key components of your direct-mail planning. One common planning tool is called CAST, for "comparative analysis of sales tools." By creating a CAST chart you can judge the effectiveness of different advertising media.

Figure 2-1 shows a CAST chart created in Excel. Here, several different possible promotional tools are compared. You can add as many of your own promotional tools as you wish. Each tool is rated on several criteria and given a value between 1 and 5. A rating of 1 marks it as very ineffective, 5 as very effective. The criteria here include:

Impact. How great an impression will it make on the viewer? A television advertisement during the Olympics could make a huge impression. Sponsoring the local ski team would result in less of an impression.

Cost per contact. Estimate the costs to use a particular promotional tool and the potential audience. A trade show, for instance, might cost $10,000 and reach an audience of about 5000 people. The cost per contact would be $2. Having a company representative contact those 5000 people might cost $100,000, at a cost per contact of $20. The trade show therefore rates higher on cost per contact.

Flexibility. How readily can your message be changed? A radio script can be changed in minutes; other tools may not be changed for weeks.

Testability. If you want to know how effective your promotional dollars are, the tool must be testable. Direct mail is exceptionally testable, while public relations efforts may yield a packet of news clips but few concrete sales. A publicity event may be as successful as a direct mailing, but its success will be much more difficult to test.

Credibility. How believable is your promotional tool?

Timing. If you buy advertising space in a newspaper, you know when that ad will appear. Mailing a brochure or other direct-mail piece, on the other hand, means you can't be certain when it will arrive in the hands of its recipients.

Closing sales. How effectively does the tool lead to completed sales?

You can add other criteria—the ease of making repeated contacts, the tool's strength in generating leads, its complexity from a management standpoint, whatever seems appropriate. When completed, an Excel CAST chart might look like Figure 2-1.

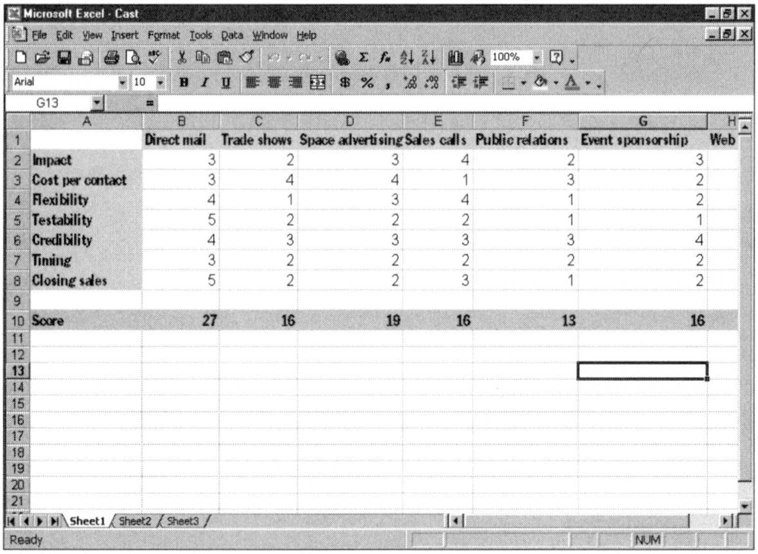

	Direct mail	Trade shows	Space advertising	Sales calls	Public relations	Event sponsorship	Web
Impact	3	2	3	4	2	3	
Cost per contact	3	4	4	1	3	2	
Flexibility	4	1	3	4	1	2	
Testability	5	2	2	2	1	1	
Credibility	4	3	3	3	3	4	
Timing	3	2	2	2	2	2	
Closing sales	5	2	2	3	1	2	
Score	27	16	19	16	13	16	

Figure 2-1

Promotion tools compared in an Excel CAST chart.

It's easy to do. Simply start Excel and open a new worksheet. Beginning in cell B2, type in the names of your promotional tools using cells C2, D2, E2, and so on. To see the entire tool title, drag the boundary on the right side of a column heading until the column is the width you want.

Then, in cells A2 through A8 (or however many you need), type in the measuring criteria. Again, you can change the column size to accommodate the label you're using.

After that, type in a number that corresponds to your estimation of a tool's effectiveness in meeting each measuring criterion. Remember, 5 is for very effective, 1 for very ineffective.

Now, to add your columns, start by clicking the cell in which you want the sum to appear, say B10, to add the column under "Direct Mail" in my example. That cell address will appear in the Name box on the formula bar, the row just above your worksheet. Next, click the equal sign (=) just to the right. To the left of it, the word *Sum* will appear. To add a column, click Sum. Now a new dialog box appears, asking you to enter the cell range. You can type it in directly (B2:B8). Or, click cells B2 and B8, and Excel will assume you want to add the entire range, as shown in Figure 2-2. Then press Enter. You'll see the sum appear in B10. That's it! Change any number in the cells, and the sum will be automatically updated.

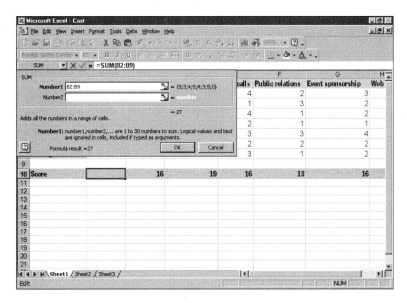

Figure 2-2

Using Excel to sum a range of figures.

Creating a CAST spreadsheet is a good introduction to the power and flexibility of Excel. Later we'll use Excel again to calculate costs of our direct-mail campaign.

Putting It Together

Schwartz Brothers Restaurants is a Seattle-area chain of seven restaurants and delicatessens. Each winter the company launches a direct-mail campaign aimed at generating sales and increasing restaurant traffic. "The first part of the year is very slow in the restaurant business," says Nancy Carlstrom, coordinator for Schwartz Brothers' direct-mail efforts. "We use direct mail to get the word out about our restaurants and give a little something back to customers who visit us during the rest of the year."

With that objective, for the 1999 campaign, Carlstrom and her creative team worked hard to ensure that the mailing (which cost several dollars per copy) gave the restaurants a worthwhile business boost. To ensure that the mailing hit the right prospects, for instance, Carlstrom combined a house list compiled from customers who had filled out in-store evaluation cards with two lists that targeted income and age levels. "We don't just blanket zip codes," she says. "Around here, you can cross a line into a zip code and get into a whole new demographic group. We want to be careful about that—if we send mailers to people who can't afford to take advantage of it, we hear back from them." The combined lists provided 167,000 households that looked promising.

The next step was to develop the right offer. That was tricky, says Carlstrom. Schwartz Brothers' restaurants range from Daniel's Broiler, a high-end steakhouse on the top floor of a 22-story tower in downtown Bellevue, just east of Seattle, to Schwartz Brothers Restaurant Delicatessens, which sell more casual, burger-oriented fare to mall shoppers. "A blanket discount wouldn't work," says Carlstrom. "It would be too steep a discount at some restaurants, while at Daniel's it would be like getting a cocktail for free." Carlstrom devised an offer for a specific dollar-amount discount at each restaurant, from $15 off at Daniel's Broiler and Chandler's Crabhouse, a premier seafood eatery, down to $4 off at the delicatessens.

Although not huge discounts, the $4 to $15 savings offer is the most that Schwartz Brothers could afford in the competitive, thin-margin Seattle restaurant scene. To make it sound larger, copy for the finished mailer touted savings of up to $66, a serious sum. And the specificity of the offer helped make it more involving than a percentage discount, which can be a moving target in a restaurant, depending on what a patron finally orders.

"The dollar figure is more concrete than a percent-off," Carlstrom says. At Spazzo, a Spanish-influenced restaurant, the $10 discount offered would purchase several popular appetizer trays. "People say, 'That's three tapas dishes—all I have to buy is the entrée,'" says Carlstrom.

For the actual mailing, Carlstrom's creative team combined two envelope inserts. One, printed on heavy stock, contained the actual offer, a business-card-sized piece that the recipient could stick in a wallet or handbag and use for spur-of-the-moment dining. The other was a fold-out brochure containing a little history on the chain, a note from John Schwartz, the president and CEO of the company, descriptions and locations of the different restaurants, and a recipe for the crab dip served at Chandler's Crabhouse. It was a classic direct-mail piece, with the offer kept simple and distinct from the more informational brochure (more on this in Chapters 4 and 5). It also was an appealing package, arriving at households in a square 5" × 5" envelope with an inviting splash of orange, orange-yellow, and lavender across the cover—a warm and inviting presentation that arrived during one of the wettest Januaries in Seattle history. Even that attractive envelope did some hard selling, though. Emblazoned across the bottom of it was the phrase "$66 of FREE dining inside." "*Free*," notes Carlstrom, "is still a very powerful term."

The response was well worth the investment made in time and money. With a 6 percent response rate, carefully tracked through Schwartz Brothers' restaurants, the mailer came close to being a direct-mail "home run," and certainly was a solid triple. And the campaign did more than get people into the company's restaurants to take advantage of the discount offer. A mention of the company's catering service in the story brochure, for instance, has resulted in a surge of calls for catering. "It has helped our whole business," says Carlstrom.

The payback from the campaign was no accident. It was the product of careful planning, an attractive offer, a well-planned mailing list, and a provocative and appealing mailing package. By taking those same steps, almost any small business can succeed with its own direct-mail campaign.

Chapter 3

Using Mailing Lists Effectively

Early in 1999, Cambridge Soundworks rolled out its new Model 88 radio, a $199 high-quality table-top radio touted by the Massachusetts-based company as the best on the market. Cambridge Soundworks obtains most of its sales via direct marketing, using catalogs, direct-response ads in magazines, and classic direct-mail packages. For the Model 88, one of the pieces mailed out was a full-color broadsheet that detailed the benefits of the Model 88 (terrific sound, small size), its features (sensitive tuner, built-in subwoofer), and the offer (special introductory price).

The full-color, double-sided mailer was not cheap to produce. Guy Daniels, vice president of marketing for Cambridge Soundworks, wanted to make certain it got in the hands of the most qualified prospects. He started with Cambridge Soundworks' "house list"—the names of people who had purchased from the company before or had requested catalogs. To that he added lists that included people who had purchased gifts from catalogs and people who had shown an interest in sound and entertainment—people such as Los Angeles Philharmonic ticket buyers. Daniels also looked for a high-end, well-educated demographic for this fairly pricey table radio, so he added a collection of *Wall Street Journal* subscribers to the mix. The response to the 100,000 pieces Cambridge Soundworks mailed out? "It was a good mailing for us," says Daniels.

Using lists effectively can make the difference between a direct-mail "home run" and a campaign that results in few sales. By choosing the right lists and using them effectively, a business can ensure that its valuable direct-mail piece ends up in the hands of as many potential buyers as possible. Indeed, direct mail is just that—it's a mailing sent *directly* to customers who a business believes should be or could be interested in its product or service. When done properly, direct mail is not the nuisance it sometimes is depicted to be. It becomes a service, one that the customer sees as almost clairvoyant in its ability to anticipate what the recipient might need. Otherwise, it's junk mail—something someone didn't ask for, shilling a product someone doesn't want and will never buy.

Few small businesses, though, take full advantage of the available technology and sophisticated customer information. Instead, they opt for simple mailings based on zip codes or local business lists. By taking full advantage of the list wizardry that's available, a small business can compete with—and beat—the big fish in the direct-mail ocean.

The Best List in the World

If you run a small business that frequently deals face-to-face with customers, you probably recognize the people who walk into your store or the clients you have met when making sales calls. You might know their first names, but how much real information about them do you have at your fingertips? Regular customers or the people you see should be the basis for your "house list," the list built around the people you know the best. When assembled properly and kept current, a house list by itself can be an exceptionally powerful marketing tool. It also gives you insight into potential customers so that you can take better advantage of rental lists as you expand your business to include markets you don't now reach.

There's another advantage to a house list: The people on it know *you*. They've already made a decision to purchase goods or services from you, and they were satisfied or pleased with the experience (you hope). When they receive a mailing with your business name on it, they are far less apt to throw it out. They may even be happy to hear from you, pleased to be singled out for attention (in this day of one-size-fits-all warehouse retailing), and even a little hopeful that their patronage has earned them a reward—a discount or other sort of special deal.

How to Build a House List

The key to building a good house list is diligence. It takes work, and most entrepreneurs are too busy keeping up with the day-to-day pace of business to make laborious data entries. But it's important that you do so. A house list is only as valuable as the data it contains—simply collecting a list of names won't do much good.

To start, you have to think of as many ways as possible to encourage people to leave information about themselves with you. Some techniques might include these:

- Put a guest book in a prominent place in your store or shop and encourage people to add their names, addresses, and perhaps notes about particular interests.

- Capture names and addresses from checks people use to buy merchandise.

- Hold monthly contests for a gift—a free meal, a $20 coupon, something big enough to pique visitors' interest. Select each winner through a drawing, and ask people to drop business cards into a fishbowl to take part. Then, using a program such as Corex Technologies' CardScan and a scanner, import the card data into your database.

- Input data as customers purchase from you. If you own a retail store but do not rely on pushing a great number of people through a check-out line quickly, perhaps you can ask customers for their phone numbers, names, and addresses at the time of a sale. By keeping careful track of each purchase, you might also be able to predict what items might be of most interest to them; you could then assemble highly targeted mailings based on customers' positions in the buying cycle for your goods or services.

- Print postage-paid response cards and drop them in customers' packages. Ask them about their buying habits, what they'd like to see in your store, how they heard about you (word of mouth, drop-in, print advertising, direct-mail advertising), and whether they'd mind receiving updates on special sales. And, of course, ask them for their names and addresses. Make them feel that by filling out the card, they'll be part of a club.

Technological tools for building a house list

True, list building takes time, and for most small businesses that's a scarce commodity. But picture the result if you add only 20 names to your list each week. In a year you'll have more than 1000 names—more than enough for a worthwhile house-list-only mailing or to use as a basis for renting and testing much larger lists. Several software programs simplify the data-capture process. Among them are these:

- **Microsoft Outlook.** For smaller lists (fewer than 1000 customers, for instance), Outlook offers enough list-making potential to make your house list walk the dog and do handsprings. You can capture names, addresses, and telephone numbers, of course, and for each individual listed you can fill in specialized information such as the customer's job title and profession and the names of superiors or subordinates. You can add any general data about the customer in a designated field, and you can flag an entry for a reminder to follow up with a mailing.

- **Microsoft Access.** For larger databases and more sophisticated data-capture tools, Microsoft Access is a powerful and easy-to-use database solution. In addition to the form fields provided in Outlook, Access allows analysis of customer profiles and offers a wider variety of reminder options and an easy-to-use interface with Microsoft Excel for data manipulation.

- **Symantec's ACT! and Goldmine Software Corp.'s Goldmine.** These contact-management applications are both popular small-business customer-tracking solutions that give you all the tools you need to build a comprehensive and useful direct-mail house list.

Keeping the house list current

List expert Paul Theroux tells of long patronizing a dry cleaning company located near his office but 15 miles from his home. "My wife always beat me up for going that far to the cleaners, but they knew my name and gave good service," he says. "Finally, I went local. But I kept getting mailings from the old cleaners, coupons and the like. They knew I wasn't coming in any more because their computer records showed this. I started to wonder what kind of system they had to take my name off their list."

Lists become old. People move (14 percent of them each year, according to some studies), their daily patterns of life change, or they may simply decide to try a different store or service provider through no real fault of yours. Do what's possible to ensure that the house list stays fresh and current. Set up a data field in the list that records the date of each purchase or other contact. If you've had no contact with a customer for six months to two years—depending on the usual frequency at which people purchase from or visit your company (a video rental store, for instance, would likely have a different "purchase cycle" than a VCR repair store)—you might consider bumping that customer off your active list and onto a list of "inactive" customers.

Don't give up, though. After all, it's easier to lure an old customer back than to find a new one. You might set up a "tickler" in your database, for instance, so that any customer who has not visited your store or ordered from you in the past six months *automatically* receives a postcard. It may say something like this: "Hi! Haven't seen you in a while. We value your business, and we hope you plan to continue as a customer. In fact, we'd so appreciate a visit from you that your next purchase will qualify for a 19 percent discount! Just bring in this card to get your discount, and think of it as a gift from us to show our appreciation for your patronage."

You can do other things as well to keep your mailing list fresh. You might send a mailing that simply asks people if they're still interested in your products or in hearing from you again. Use a prepaid self-mailer so they can easily respond; anyone who takes the trouble to acknowledge receipt of the card and check the box that says "Yes! I'm still here and would like to continue receiving information from you" is worthy of another mailing soon afterward that extends a special offer.

And don't forget the telephone. This can be a touchy subject, given the volume of telemarketing calls people receive. But if your house list is manageable, checking by phone with customers who have not made contact for a year is probably worthwhile. It may mean making several calls a day, but if you don't let the numbers pile up, the task can be manageable and quick. Again, use Access, ACT!, Goldmine, or another contact manager to remind you when a customer listed in it hasn't been in touch for some time.

Rental Lists

House lists are terrific but, of course, a bit limiting. They catch the fish within the radius of your ability to throw a net. What if you could hire someone to throw a really big net, much farther than you could, with special gaps in the mesh to catch a certain kind of fish? That's the basic concept behind renting lists.

The idea of renting a mailing list may be a bit intimidating at first. There are thousands of lists from which to choose, which alone can make it hard to decide what list or lists to try. Most lists are available for rent only in lots of 5000 or more; that number adds up quickly when you realize that a good test may require that you rent as many as 10 different lists. But the costs for renting a list are not astronomical—typically between $50 and $100 per thousand names—and you do not need to use all the names the moment the list is available. Even for mailings of as few as 1000 pieces, renting a list may well make good business sense.

Compiled Lists

Pick up your phone book. If you were to prepare a mailing based on all the addresses and business names in the yellow pages under, say, "Automobiles—repair and service," you would be using a compiled list. A compiled list lumps people together by business, city of residence, or occupation. Compiled lists are taken from phone books, motor-vehicle records, city directories, credit reports, association directories—any record where people leave behind a name, address, or other information. Compiled lists in aggregate probably contain nearly every adult human being in the United States and most of their youngsters, so if you have big plans for a mailing, they might be right for you.

The downside to a compiled list is that you don't know much about the people on it aside from the fact that they all, say, live in New Orleans. You don't know, for instance, if they are apt to respond to a direct-mail solicitation, how much they spend on average, or the type of offer to which they respond. Using compiled lists is often like using a shotgun, not a rifle. The chance that you'll waste a great deal of postage is high.

Still, compiled lists are not devoid of useful information for the direct mailer, says David Ohringer, a list broker and direct-mail packager in the Los Angeles area. He offers the hypothetical example of a business selling wrought-iron security fences. The owner knows from reading the newspaper that a spate of break-ins has occurred in a particular part of town. Using a compiled list, that business owner might be able to send a mailer to all the homes in the zip code where the burglaries are occurring. A better approach would be to use a compiled list but to filter it based on home price and length of residence (both pieces of information that are likely to be available). The result: You might target homes that are fairly high in value and whose owners have built up moderate to substantial equity. Those people would have more to protect, and they would have sufficient equity to allow them to purchase wrought-iron fencing, a high-ticket item.

Compiled lists also can be broken down by what is called the standard industrial code (SIC). In a compiled list, for instance, the SIC for an automotive repair shop may be 7549. Nationally or regionally, you could find other shops that do similar work based on that SIC number, perhaps preparing a mailing to offer them a special tool or service that's particular to auto repair. You also can find shops that specialize in other types of automotive repair, such as brake repair (7538) or Saab-only work (55310001), each of which might be a profitable niche market for your goods.

Response Lists

If you sell goods or services to the general public, response lists are your best choice. Think of them as the "house lists" of other direct-mail or direct-response businesses. That is, in fact, what they are; response lists are sold or rented by companies as additional revenue sources. They are lists of their best customers—people who have ordered via direct mail in the past or have shown an interest in a product such as yours. For direct mail in particular, studies show that people who have responded to such appeals in the past are more likely to do so again than those who have never ordered via direct mail. That's an important point if you hope to generate most of your business via direct mail. Even if your mailing effectively targets people who should be interested in your product, if they aren't favorably disposed to direct-mail offers, your seemingly well-aimed pitch will go nowhere.

Where to Find the List That's Right for You

You can begin finding response lists by looking through the SRDS Direct Marketing List Source. SRDS (Standard Rate and Data Service) is the industry-leading source of data on everything from advertising rates in magazines to response and compiled lists. Lists are, well, listed by both consumer and business categories. You can find the SRDS directory in most libraries. In addition, the company runs a comprehensive Web site at *www.srds.com*, which includes a sample list. In it, you'll find such information as the following:

- The name of the list broker or manager
- A description of who is on the list (for example, "buyers of interior furnishings for commercial and residential use")

- How the list can be selected (by zip code, state, carrier routes)
- A breakdown of the job titles on the list
- Commission and credit policy
- Labeling options
- Rental rates and delivery schedule

You'll find hundreds of lists of all sizes and varieties as you look through the SRDS directory. You're almost certain to see dozens of potential markets you hadn't thought of before. Some entrepreneurs, in fact, develop products and offers solely through inspiration provided by lists. They find a market, figure out what its unmet needs might be, and then do a mailing to test their concept for a new product or service.

What Can a List Broker Do for You?

If you find a list—or several lists—in SRDS that seems right for you, there's nothing to stop you from calling the manager of that list (contact information will be shown on the page) and negotiating a rental. But for most businesses—particularly those new to direct mail—working from the SRDS list is at best a good starting point. The amount of information is overwhelming, and much of it is apt to be meaningless. Sources such as SRDS can lead you to the list manager (a business that acts as a list retailer, renting out lists that it "stocks"), but that manager is not likely to tell you if a better list for you is available from another manager.

This is where the list broker comes in. A broker is an independent list agent—he or she receives a commission from the list manager or owner but otherwise owes no particular allegiance to a particular list. Nor will a list broker hide some list "bargain" from you; lists are published on a fair-trade basis, and it is in the broker's interest to get the best possible list in your hands. Says Mary Kerford, a direct-mail advertising executive with the Cohn & Wells agency in San Francisco, "We absolutely advocate working with brokers. We have a good idea of what lists are out there, but brokers are on the cutting edge and know about all the lists coming on the market. For instance, when the list for *PC Week* started to come out on a quarterly basis, we let the broker know that we wanted it the moment it came out. He kept his eye on that for us while we were able to do other things with our clients."

Brokers can offer other helpful services. They are in the direct-mail business, so they may be able to recommend to you writers or designers to help with your project. They may be able to help you with testing and additional marketing possibilities. Brokers also have at their disposal special profiling and modeling tools to help evaluate lists and extract the most useful lists from the thousands available. Brokers can look for lists based on such characteristics as these:

Past purchases. If your goal is to sell furniture for a home office, it would be useful to know who has bought products in the past that have a home-office orientation. Some catalog and mass merchandisers can break down their lists based on buying habits, allowing you to rent lists that include people who have bought such equipment in the past or who subscribe to *Home Office Computing* magazine. A good broker can look at your product or service and help determine which so-called "affinity" lists might work best for you.

Location and demographics. You may sell items that are likely to appeal to people of a certain income bracket. Lists often can be built along demographic lines, so you don't offend people by sending mailers for items they couldn't possibly afford. Similarly, you can build a list to target people who live in certain locations.

Recency, frequency, monetary value. This "RMF" formula is something of a mantra among many direct-mail experts. It is based on how recently someone purchased through direct-mail channels, the frequency with which that person makes such purchases, and the monetary value, on average, of those purchases. Ideally, you want to target your mailing to people who shop early via direct mail, shop often via direct mail, and do not hesitate to spend fairly large sums of money. Some lists may also indicate how people made their purchases (with a check, with a credit card, and so on). Lists with "hotline" buyers—people who have made recent purchases via direct mail—will often cost you twice as much. But those names may well be worth the extra money.

Universe. List experts talk about the "universe" of the list, or how big it is. The universe can indicate how likely the list is to contain your possible mail targets. Suppose, for instance, that you want to reach people who spend an average of $500 a year or more on gardening supplies, and you target that list further by identifying certain zip codes where you know gardening is a popular pastime. If a potential list offers only 20,000 names from all across the United States, you can assume it will yield relatively few candidates for your mailing.

Psychographics. This term refers to the habits, attitudes, and behavior patterns of individuals. If you are marketing upscale travel-related services, you might want to look for people with a certain income level and whose children are grown, so they have more time and money on their hands. You might also look for purchase patterns or magazine subscription lists. The readers of *Backpacker* magazine, for instance, would be better candidates for travel that emphasizes adventure and a certain rough-and-ready nature than the readers of *Travel & Leisure*.

Working with a list broker

To benefit as much as possible from a list broker's services, you should follow a few guidelines. For starters, talk with several brokers. For a small business a good broker can be an invaluable source of information and advice. Look for one as if you were looking for a bank officer or a real estate agent—ideally, someone you can talk with, whom you feel free to ask questions, and who seems genuinely interested in your success. If you have the budget, you should test lists from several brokers, and then "roll out"— execute the complete mailing—with the list or lists that work best.

David Ohringer suggests that you give the broker all the help that you can so that he or she can work with you more effectively. He advises first that you be open about the product. Some clients are afraid that a broker will take their idea and rush immediately to the patent office, says Ohringer. "But if we don't know what it is, it's next to impossible for us to find the right list," he says.

Also, mailing novices should take the time before contacting a broker to figure out what they can afford to spend. "People call me up and say, 'I want to do a mailing. How much will it cost?'" Ohringer says. "But there are dozens of variables. It's not like going to the store and buying a Hershey bar." He also suggests that you be honest about your budget and give the broker a concrete figure for what you plan to spend.

Renting a list

If you work with a broker, the process of renting a list is straightforward. Depending on your budget and what lists seem promising, your broker will place an order for the lists with his or her managers—companies that collect lists from list owners (magazine subscription departments, for instance)—and prepare them for rental. As mentioned earlier, costs vary, but most lists run between $50 and $100 for every 1000 names. When the order is placed, most managers will ask to see either a finished mailing or an accurate mock-up of the mailing. The reason? They want to ensure that you don't compete directly with a company that sells its list to the manager.

Your list can come to you in several different physical packages. Magnetic tape is preferred because it gives you the greatest flexibility. You can have envelopes specially typeset, for instance, or input the list names into a salutation in your letter. Alternatives include Cheshire labels, which are printed on computer paper and then glued to envelopes, and pressure-sensitive peel-off labels ("P/S labels" in industry jargon), which can be removed from the envelope and stuck onto your response card. Your broker can advise you on the best labeling type for your mailing.

Remember, you are *renting*—not buying—the names on that list. Once you have your lists in hand and are still smarting a bit over what they finally cost you, you may be tempted to surreptitiously reuse the list one or more times. Don't. List managers all plant dummy names on the list, and mailings to the dummy addresses end up on list managers' desks. If they get multiple mailings to these fake names, they'll know you're cheating. At that point, you've effectively been caught stealing.

Once people on the list begin responding to your offer, however, they become *your* customers. Then you can add them to your house list and send mail to them repeatedly. Also, you are not obligated to use every name on the list you rent; remember, most lists come with a minimum order of 5000 names. You might, for instance, use only half the names on a list or combine 25 percent of the names on each of several different lists. In these cases, you can use the rest of the names for a later mailing. If you try one of these approaches, be sure to randomize the lists by using every second, third, or fourth name; don't simply use the first or last 1000 names on a list. And you don't want to let a list sit on the shelf too long. Even the best lists are probably only 85 percent accurate in terms of an actual person still living at the address claimed. Within six months the accuracy rate could drop markedly, and it will continue to deteriorate. A list more than a year old probably is not worth using.

It's also wise to combine several lists rather than using all or half of one or two lists. Each list is apt to result in a different response, some good, some not so good. By using portions of several lists you increase the chances that at least one of the lists will prove successful. Then you can mail to the rest of the list or rent additional lists that resemble the successful list. True, when you're spending $75 to $100 for each set of 1000 names it may be tempting to cut costs by renting only one or two lists. Don't do it—budget for several lists, even if that means cutting back on the fun stuff, the "creative" package with the letter and brochure. A decent mailing in the right hands will likely do well. A great mailing in the wrong hands won't accomplish a thing.

Database Mining

"Database mining" is a techno-geek term that is getting a lot of play in the direct-mail world. Essentially, the term applies to one of several methods of making the most of a customer list, either a house list or a rented list.

A good database miner can start with his or her house list. As discussed earlier, good records on your customers allow you either to target them with more specific mailings or to leverage your house list into a highly targeted response or compiled list. The basics include knowing the following:

- How recently the customer purchased from you.

- *What* the customer purchased, as specifically as possible. Don't simply lump together purchases under "clothing" or "merchandise." If somebody bought a charcoal gray suit in size 44 long from Hickey Freeman, make a record of it.

- How often the customer has purchased from you since the first contact with you.

- What *first* led the customer to visit or contact you (word of mouth, direct-mail piece, newspaper article, just walked past).

- The number of people in the household or business.

- If this is a business customer, his or her title, the sales volume of the company, and the number of employees.

Once you understand who your customer is, you can do several things. If you run a small business and are working primarily from your house list, for instance, you can execute highly targeted mailings. Let's say you run a tool shop that caters to contractors and high-end do-it-yourselfers. You know that one set of customers is partial to DeWalt power tools, another to Makita. If you happen to be overstocked on DeWalt tools or if you add a new sliding compound miter saw to your inventory, you can send out mailings offering discounts on or accessories for those particular brands. With a really good database, you might even know what tools a customer lacks; you can then invite him or her to come in and take a look at the tools he or she does not yet have.

By knowing who your customers are, you also can figure out who might make good new customers. That house list can be "overlaid" with various compiled lists, each of which adds to a piece of data you have on the house list. As the owner of that tool shop, for instance, you might look for subscribers to *Fine Woodworking* magazine, and then sift through that list for individuals who resemble your typical customer in age, income level, and home ownership. They would likely be hobbyists, many retired, with considerable disposable income. Their interests would run to versatile all-in-one shop tools and sets of high-quality wood chisels. Another set of your customers might resemble the readers of *Custom Home*. They'd likely be more interested in rugged worksite tools that offer good value—framing hammers, but not wood-boxed chisel sets. By "mining" your house list, you can get a good idea of how people who appear on rented lists do or do not resemble your current customers, and you can build mailings that target these new customers accurately.

Other List Tools

The proverbial kitchen table—with a PC on it—is for some entrepreneurs probably a fine place to begin a direct-mail campaign. And several software products make the task far easier than it formerly was.

Many mailings, for instance, could be assembled using DeLorme's popular software suite of Phone Search USA and Street Atlas USA. Using Phone Search, a mailer has access to a compiled list with more than 100 million names. Searches can be executed based on business type, zip code, or SIC code and then exported to a contact manager such as Microsoft Small Business Customer Manager in Office 2000 or a database such as Access 2000. Addresses also can be printed directly onto Avery labels for smaller mailings.

Another alternative is a software program called MarketPlace, distributed by iMarket, Inc. Early in the 1990s iMarket acquired the rights to the business-to-business CD-ROM database technology from Lotus Development Corporation. Its reengineered product contains a marketing database of more than 10 million U.S. businesses. With MarketPlace, customers can perform surprisingly sophisticated market research that in many cases enables an entrepreneur to develop his or her own lists.

41

One iMarket customer is Quick Fuel, Inc., operator of a network of unattended, automated fuel stations for trucking companies. Through a computer system, customers use a magnetic card to obtain diesel fuel or gasoline at any of the 37 stations located throughout the Midwest and southern United States.

To identify potential customers in new markets, Quick Fuel uses iMarket to search for prospects with profiles similar to those of their existing customers—mostly regional trucking companies with 10 or more employees. The company plots those names on a map to check their proximity to Quick Fuel fueling stations and sends mailings to the most likely prospects. Quick Fuel purchases the desired list from iMarket, and it is available for repeated use for one year. "It's really allowed us to target our best prospects and cut down on mailing costs," says Chad Dombroski, project manager for Quick Fuel.

The Bottom Line: Did It Work?

No list is perfect. You'll inevitably find nonmailable addresses—"nixies"— on any list. You may find that several lists result in poor responses. And lists often are easy targets for a new mailer's wrath, says Roy Schwedelson, president and founder of Worldata, a prominent list broker and manager. "Picking lists is boring, so most entrepreneurs don't participate in that," he says. "Instead, they're involved with the so-called creative (writing and design), and because they've worked on that, they never think there's anything wrong with the creative—it's always the list's fault."

"But," continues Schwedelson. "Even if the list wasn't 100 percent clean, the question is: Did you learn from it? Was it successful for you? If you tested 10 lists, don't expect 9 to work. Three would be terrific because now you can roll out with those 3."

Chapter 4

Direct-Mail Copy That Sells the Goods

"When it comes to retirement, don't get caught short."

"You're in! So get ready for more!"

"You've earned it!"

"Homeless children freezing."

Those phrases all come from recent direct-mail campaigns, several of them award-winners in the ECHO competition, the annual awards competition of the Direct Marketing Association. More importantly, these campaigns all got results—response rates far above the usual 2 percent, along with money-making sales. Even though they are only headlines, they all highlight important lessons about writing copy that really draws recipients in and moves them to act.

Direct Mail—The Word Is the Message

Most people think of "advertising" as what they see on television or in a magazine. They think that advertising means selling beer with talking frogs or cars with vistas of sinuous roads or home loans with a pitch by an NFL quarterback.

If that's what *you* think advertising does, you're right. That's "advertising." "Direct mail" is an entirely different kind of creature. You aren't trying to create warm, fuzzy, or even sexy thoughts about your product or service. You're trying to get people to buy it. It is that distinction that fundamentally separates direct-mail marketing from advertising.

And how do you get people to respond to direct mail? With good copy. Look through a selection of direct-mail pieces. You'll find them full of words: about the offer, about the advantages of the product, about the benefits users will enjoy, about the experiences of other people who have used the product or service. Many direct-mail pieces, in fact, rely on words for 75 percent or more of their content, a far higher proportion than in soft, image-building advertising. "Copy is the prime method we use to communicate," says David Withers, a direct-mail copywriter and designer based in Pasadena, California. "Pictures and graphics add to it, but our main tool is language."

On one hand, that's a tremendous advantage for the direct-mail marketer. You don't have to pay a film crew to bathe your product in gauzy light and pose it with Big Sur in the background. On the other hand, you have to come up with credible, persuasive, readable copy. And good writing is hard to come by. All those sticky rules of grammar. All those phrases that emerge on a computer screen that sound nothing like the real language of real people. All those smart-guy readers who look at a piece of work still damp from the writer's sweat drops and yell, "Who cares!?!" It's no wonder that too many direct-mail novices fall back on cues they pick up from the world of mass-market advertising and give short shrift to good copy. With a clear idea of what it is you want to do, some useful tips, and a little (OK, maybe a lot) of effort, you can turn out solid direct-mail copy, too. Or know what *is* good copy when a subcontractor produces it for you.

Focus on the Customer

Here's a rule to remember: Households don't buy things. Families don't buy things. Companies don't buy things. Company departments don't buy things. *People* buy things. And people are more or less all alike. They have problems to solve, bosses to please, underlings to keep in line, families to take care of, cars that need their oil changed, dogs that chew up the curtains. If there is one key to writing successful direct-mail copy, it is this: Write as if you were writing a letter to another human being. That simple step will yield multiple benefits. Your writing will probably sound more natural and conversational, which is good. It also will flow more easily, as you won't be all knotted up about writing to a corporate executive or people of a certain "type." And you'll be better able to tap into the key motivators that lead people to buy something.

Now for a bit of a letdown: It's almost a sure bet that one of those motivators will *not* be the fact that you've labored long and hard to bring this wonderful product or service to market and that its benefits will be obvious. As direct-mail ace Bob Hacker says, "Nobody cares about you. They only care about themselves." So don't try to get a customer to buy your product because they admire you, like you, or even feel sorry for you. They won't. They *will* buy your product, however, for any of these reasons: fear, greed, exclusivity, guilt, or flattery.

Fear

Fear is the motivator of the nineties, says Hacker. "Everything has been going great—the stock market is up, and the economy is strong," he says. "But at the same time there are all these layoffs and mergers and downsizing. So people are afraid." One surefire way to tap into that fear: Imply that other people are taking advantage of your offer and are going to best the reader as a result. In the early 1970s, for instance, the *Wall Street Journal* launched a direct-mail promotion known as the "two men" campaign. To paraphrase, it tells the tale of two men with similar backgrounds who began their career with the same company and with approximately the same job. Over the years, one enjoyed a succession of career advances, while the other languished. Why the difference? One read the *Wall Street Journal*. The other did not. "That ad has never been beaten," says Hacker. "It uses the 'don't be left behind!' formula perfectly." Fear may remain a successful pitch for some years to come, as the massive Baby Boomer generation reaches the years when its members will be increasingly nervous about health, retirement, and finances. One of the headlines at the start of this chapter—"When it comes to retirement, don't get caught short"—appealed to the fear many people have that they will not have enough money for retirement. It was from the investment firm of A. G. Edwards.

To make fear work, though, you need to present your product or service as a solution. Your pitch should show how your product could help build wealth, improve health, or lead to career advancement. The direct-mail piece from A. G. Edwards, for instance, invited the recipient to get a retirement income analysis from an A. G. Edwards financial advisor. And the pitch should be simple and easily understood—as was the *Wall Street Journal's* "read the *Journal* and succeed" campaign.

Greed

Greed was big in the 1980s, faded a bit as an appeal in the 1990s, but remains a reliable way to lure the buyer. Put more delicately, any ad that focuses on saving or making money for the recipient appeals to greed. It is a staple with mutual fund appeals, of course, and it is the basis for any campaign that emphasizes a special price or sale. In fact, any good offer appeals to the human desire to save or acquire money and material things. Greed, though, is a double-edged sword. Too much emphasis on price will leave you with a clientele made up of bargain hunters who will abandon you the instant they can get a better deal elsewhere. Be especially wary of offering gifts to customers solely for visiting your store or showroom. You'll find people dashing into your store, asking for their free gift, and dashing back out again before you have a chance to make a pitch.

Exclusivity

You remember one of the headlines that opened this chapter: "You're in! So get ready for more!" It appeals to another human desire—the desire to feel as if you are part of a particular club or have been given special status. There are many ways to suggest that the recipient of a direct-mail piece has been carefully chosen. You can say that you serve customers only in a certain geographical area, or of a certain income or educational level, or who have moved a particular distance up the corporate hierarchy.

Of course, the key to using exclusivity effectively is that your list must accurately reflect the desired target for the campaign. It won't do much good to appeal to exclusivity and have your campaign land in the mailbox of a household that can't afford to pay your product's hefty price.

Guilt

We should all do more—more volunteer work, more charitable donations, more exercise, more attendance at important cultural events. Playing to the guilt most people feel about doing more can make an effective direct-mail headline. Charitable groups in particular often play the guilt card in their direct-mail solicitations; the headline "Homeless children freezing," at the start of this chapter, was from a campaign to raise funds for Covenant House, a New York shelter for children. It made a direct appeal to sympathy. The pitch can also be indirect, such as sending "free" tokens such as holiday stamps or note cards and then asking for a donation.

Flattery

Few people possess genuine self-esteem. As a result, it's often possible to flatter them into buying a product—or at least into thinking about making a buy. Remember the headline at the start of this chapter, "You've earned it!" It was from a recent American Express direct-mail piece aimed at holders of the company's "green" card. The goal: to upgrade people to the "gold" card, which has few tangible benefits for its holders and costs more to use, but is marketed by American Express as a more exclusive card. To encourage the upgrade, American Express congratulated green card holders for having timely payment habits and other appealing traits and dangled the gold card as a reward. Transparent but effective—response rates ran strong early in the campaign.

As with the use of exclusivity, deploying flattery achieves more success if you know the target audience well enough to give the commendation a ring of truth. American Express, for instance, used its house list of current cardholders and did indeed know enough about their credit habits to make a reasonable appeal to vanity. With good information, you may be able to compliment purchasers of a particular clothing line on their good taste, to praise readers of environmentally oriented magazines such as *Audubon* or *National Wildlife* for their appreciation of the natural world, and to stroke the intellectual egos of Stanford or Harvard alumni or book-club buyers of Wittgenstein collections.

Turn Customer Motivation into Action

Now that you know what might move recipients of your direct-mail piece to purchase your new high-speed hard-disk defibrillator or visit your just-opened Lapland-theme restaurant, how do you get them to act? Here's where good copy comes into play. A well-written story for your product, one that accurately yet aggressively promotes its features while making a conscious appeal to one of the motivators just listed, should yield the results you desire: sales.

Don't just start banging away at the keyboard. Before you begin to write, you should fill several sheets of paper, one for each of the questions that follow, with as much detail as you can muster.

Who is the customer?

Who might buy your product? What will motivate customers to do so? Are they apt to be worried about something? To desire a certain quality in their lives? To be part of a fairly select group? Try hard to visualize the customer; consider drawing on several traits of your friends or relatives that might exemplify your potential customer. Some copywriters even talk about propping a photo next to their computer of a person they see as their "typical" customer and writing to that customer.

If you can, this is a wonderful time to conduct some formal or informal surveys. If you already have a list of customers, invite 10 or 12 into your store or place of business to discuss your product. In exchange for a gift from you, or even a small cash payment, get them to talk about what moved them to make the purchase.

One objective here may sound counterintuitive. You ultimately want to market to the *smallest* group that you can, not the largest. It's too difficult to devise an offer and a marketing letter that will appeal to both, say, 25-year-olds and 65-year-olds, or small-business owners and chief information officers at Fortune 1000 companies, or bird watchers and bird hunters. Successful direct mail is targeted direct mail: It reaches a group of people who by age, income, buying habits, or place of residence are favorably disposed toward your product.

You also want to keep in mind that most people fall into one of three categories, says David Withers. One group consists of those people who are essentially desperate for your product. They'll snap up even a middling offer. For example, a driver just missed having a serious auto accident because it was raining and the car's bald tires slipped; he or she got your mailer for a tire sale, with a 10 percent discount offer, at home that day.

Another group isn't going to buy your product or service regardless of the offer or the wit in your copy. They don't drive, for instance, so tires won't do them any good. The people to whom you need to write, the third group, are those who *might be interested* in your product or service, says Withers. They're thinking about getting new tires, but they think they can get another 10,000 miles out of their existing set, or they have traditionally bought tires from a competitor of yours. "Those people are the 'maybes,'" says Withers. "If you do it right, they can be convinced to buy your product."

What's your offer?

Once you have a mental image of your customers, you can tailor your offer to best suit them. What is apt to be most appealing to that customer? A straight discount? Special financing? More information? Picture your potential customer holding your precious mailing over the trashcan. What will keep it from landing in there? Remember, with direct mail you can't stand and gauge the customer's response to your product or price and then throw in an extra 10 percent off if you seem him or her wavering. You have to take your best offer shot right now.

Keep in mind that the offer is your chief weapon in reaching wavering buyers, says Bob Hacker. For every additional 5 percent you discount a product, you may pick up another 10 or 100 or even 1000 buyers.

What are the product's benefits?

Pull out the evaluation of your product that you wrote while reading Chapter 2. On a fresh sheet of paper, list all the benefits that you think will appeal to your hypothetical customer. It is very important that you see your product from the *customer's* point of view, not yours. Nobody cares that you think it's wonderful or that you worked hard to develop it. "What people want to know is 'What's in it for me?'" says direct-mail copywriter Deborah Jason. "Will it make my teeth white? Will I lose 10 pounds? I always go over these points with my clients. But it's amazing how often people don't really know their products."

Keep in mind that product *features* are not the same as *benefits*. Bob Hacker says the difference is that between the drill and the hole. The drill may have lots of features—5000 rpm, a quick-change chuck, a comfortable grip—but the benefit is a great hole. "We always try to sell the holes," he says.

What is your objective?

What do you want the prospects to do? Buy your product? Come into the store or restaurant? Qualify themselves as interested in your services by asking for more information? Each approach may require a different technique in your copy. That isn't to say there is one *particular* approach to use for each objective. If you don't know what it is you want the customer to do, though, you won't be able to frame a letter and offer that will reach that customer. "Direct mail is too expensive to use as a passive vehicle," such as simply announcing a store opening, says David Withers. "If you don't ask for a response, then you're wasting postage."

First Drafts

Now it's time to start writing. Keep this in mind: First drafts are written to be thrown away. Don't fall into the trap of thinking your first words have to be perfect. They won't be. Writing is a *process*, not a leap from raw idea to perfection.

In fact, the first part of the writing process is simply to brainstorm as many ideas about your product and how to sell it as you can. Take out a clean sheet of paper, get yourself focused on your product, and start writing down all the reasons you can think of for buying that product. Range as far and as freely as you can in coming up with ideas. In fact, be a little outrageous.

Making a Mind Map

When most people brainstorm, they fall into a trap they learned in junior high: Right from the start, they write down a big number "1" on the sheet of paper, with maybe a "2" and a "3" after that. The minute you do that, you're sunk. You've created a hierarchy based on priority, and now you must come up with something that fits that No. 1 designation—best, biggest, first, worst, whatever. You can't do it; your brain will lock up.

Instead, make a "mind map." Take out a clean sheet of paper, and then write in the middle of it a phrase that describes your product; let's say a new kind of insulated travel mug. Draw a circle around that phrase (it can be just a few words). Next, draw a few straight lines out from that circle. On those lines, start writing down whatever comes to mind about your new mug. Don't mentally edit—write down whatever comes to mind, no matter how goofy it sounds. If one idea seems related to another you've already written, draw another line branching out from the first.

In time, you should have a circle that looks like it's having a really bad hair day. And you should have lots of ideas written on your map. This technique allows you to write ideas as they come to you—randomly, not hierarchically. Once they're on paper, you can begin to look them over to see what might work and might not work. *Then* you can start the process of picking the ones that you expect to be most effective.

Writing a Direct-Mail Letter

A typical direct-mail piece consists of several parts: the envelope, the letter, the brochure or broadside, and the response card. We'll discuss several of those elements in Chapter 5. For now, we'll focus on writing the key part of a direct-mail piece: the letter. Even a bare-bones campaign can afford that much.

And that letter is a powerful tool. It's the salesperson standing just inside the prospect's door. It's the letter's job to get the prospect interested in the merchandise. The letter also is the most fundamental component in a direct-mail package. If you have a budget of a dime per piece, you can still afford to send at least a letter.

Sometimes it is surprising how much time people will spend with a piece of direct mail once they open it. If you write a good letter that can keep the attention of the reader, four or more pages are perfectly acceptable. In fact, for "considered purchases"—items people have to think about buying, not impulse buys—long letters fulfill the need to learn as much about the product as possible. Whether you write a short letter or a long letter, the key is to make sure every word does a job and moves the reader one step closer to making a purchase.

It's also important to think about the structure of a letter. Most letters have four key components: the headline, the lead, the hook, and the body.

Write a Powerful Headline

A headline should be punchy, direct, and as informative as possible. To write a good one, focus on a single benefit of your product, proclaim your offer, cite an advantage your product enjoys over a competitor, or make a promise to the reader. Here are some examples:

- Ride the ultimate bicycle tour in Colorado (benefit bicycle ride around Colorado)

- Give your child a back-to-school boost with award-winning Edmark software! (Edmark educational software)

- Simplify your life! (Legato backup management software)

- Truly great sound is now truly simple! (Cambridge Soundworks Model 88 table radio)

- We dare you to have a good time (National Geographic's *Adventure* magazine)

Headlines with the words "how to" are also effective; they're action-oriented and tell the reader that they can achieve a goal by following your program or buying your product. Using numbers in your headlines—numbers that highlight price savings or count the number of product benefits or give a deadline ("You have 30 days left to save!")—is also effective.

If you can, write a headline that forces readers to think about their motivators. Ask them a question: "Do you want save $100 a month on heating bills?" or "Do want to live a longer, healthier life?" or "Is there a dog in the house?" If a reader responds with a "Yes!" while reading that headline, then you've begun a dialog and involved him or her in your product and its story. From that point on, the reader will be more favorably disposed toward your product than if you had not posed that question. Don't worry about "leaving people out"—cat owners, for instance. If you're selling a new kind of low-fat dog treat, you aren't going to sell any to cat owners anyway; forget about them and direct your words to the people who need your product.

Finally, avoid getting cute in your headlines. Don't write to amuse yourself or win awards—write to win sales. Focus on benefits and the customers' possible motivators. That's the formula for good headline writing.

Many direct-mail letters actually start with a boxed piece of copy above the greeting or salutation. Called by some direct marketers a "Johnson box," this opening element is used to underscore a particular benefit or simply get the reader interested in what's coming next.

Right after the headline or other opening comes the greeting. Sure, some direct marketers can afford to print a recipient's exact name in the salutation, but for most direct-mail projects that isn't possible. Lois Geller advises that you skip the fake camaraderie, however; the minute you write "Dear Friend," your reader is apt to answer "No, you're not!" End of sale. Better to keep it simple and generic: "Good afternoon" or even just "Hello!"

Draw Readers in with the Lead

Now that you've shaken the reader's hand, you can make the pitch. We'll start by writing what's called a "lead," a term borrowed from the magazine world but a highly useful model for direct mail. If you look through several magazine articles, you'll find that most begin with an anecdote that is relevant to the rest of the story. In that opening anecdote, an individual will confront a crucial decision, receive bad news, face an obstacle, or achieve a goal. By writing such an opening, the writer immediately involves the reader in an action-oriented story that involves a single person. This approach achieves two things. First, it gives the tale some forward motion, as the reader wants to see what happens next. Second, by focusing on an individual it brings the story down out of the clouds and makes it concrete and specific. Together, these two elements satisfy a very human need to hear stories and identify with particular people.

How can you use this technique in direct mail? Easy. In the case of our insulated travel mug, mentioned earlier, you might tell about the time you had settled in for a long drive, coffee cup in hand, but soon spilled the cup because it wasn't the right size for the cup holders in the car or about how you'd long wished that your morning double-tall latte stayed warm during your 30-minute commute. A letter selling financial advice might illustrate the case of a 45-year-old computer salesperson who had finally realized that investing, not striving for higher salaries, was the only way to achieve financial freedom. Keep the opening part of the letter short—a paragraph, two at the most. And make it as engaging and real as you can.

In some cases, though, you'll want to start more briskly, perhaps by announcing your offer in the first sentence and stating what a customer should do in response to it. Here's a letter Bob Hacker wrote for Airborne Express, announcing a new mailing service:

Dear Ms. Collins:

I want to give you a FREE big-display calculator, just to show you that cutting business costs doesn't always mean sacrificing service or settling for second best.

Now you can get Airborne's excellent next-day service while saving even more money on your air express shipments—with Flight-Ready prepaid express envelopes.

And for a limited time, you can order as many Flight-Ready envelopes or packs as you want and save 10% off our regular Flight-Ready price.

In three sentences, the letter's recipient is told what she'll get (a calculator), what it will cost her (nothing), and what she needs to do (order Flight-Ready envelopes at a discounted price).

Grab Readers with the Hook

Now for the rest of the story. The lead sets up the premise for your product by telling a story that engages the reader at an emotional level. It's like the lure you use when fishing—it tempts the fish to chomp down. Once a

53

fish bites, you want to set the hook. In a direct-mail piece, you do so by moving from the lead to a paragraph or more that highlights the key *benefits* of the product. Remember: benefits, not features. Holes, not drills. Say to the reader: If you've faced a problem like the one we just outlined, then this product is for you. Buy it, and you'll live longer, make more money, and win the affection of complete strangers. The hook should not be long—another paragraph, perhaps two.

Make Your Offer in the Body

Next, flesh the letter out in the body. Review the benefits, using more detail than you did in the hook. Tell other stories about other users. Make your offer. Reassure the buyer with a clear explanation of the warranty, the free-trial offer, or the money-back guarantee—whatever method you use to assuage their doubts.

Try to put yourself in the buyers' shoes. What questions are they apt to have? What objections? What do they need to know to feel more comfortable with the purchase? What is apt to motivate them, and how can you tap into that motivation? Try to imagine the order in which the reader may raise these questions, and answer them in that order.

Also, cite your offer several times. Cite it in the hook, then again in the body, and probably once more at the conclusion. Include it so much that it sounds as if you're beating it to death. People will often skim a letter, jumping from the opening to the ending, with perhaps a short landing in between. You want to make sure they see the offer. In the Airborne Express letter, for instance, Bob Hacker mentioned the 10 percent discount and free calculator four times each.

Be sure to tell the readers what you want them to do. Urge them to order while the offer lasts or to call for more information. Remember, you have to *ask for the sale*. Force readers to make a decision, to say "yes" or "no." "Maybe" won't help you; neither will "maybe later." Direct-marketing experts will all tell you that a direct-mail piece tossed into a "deal with it later" pile is dead, not dealt with.

Don't Forget the P.S.

Put in a postscript as well. Like the "Johnson box" at the start of a direct-mail letter, a P.S. is a common tool for the direct-mail copywriter. It's a chance to reiterate the offer, state once again that you have a bulletproof guarantee, and again urge the reader to take action.

How to Strengthen Your Writing

Ask many people to describe a sunset and they'll use words such as "amazing" and "spectacular" and "awe-inspiring." Well, you might use the same words to describe Pamela Lee Anderson's synthetic bustline, but obviously they're rather different things. Adjectives such as "spectacular" really don't mean much. It's tempting, though, to use lots of them when touting a product. Avoid the temptation. Use verbs and nouns—concrete, action-oriented words. Instead of saying your innovative travel mug has a "huge capacity," for instance, say that it "holds 24 ounces of coffee—the equivalent of two double-tall lattes!" That phrase also uses comparison, another useful tool as it puts an item that's perhaps new to readers into a context they already understand.

Direct-mail copywriters can take a tip from successful catalog marketers such as L.L. Bean, a company whose writers are masters at making things sound useful, practical, and appealing. Here, for example, is the description for the company's Sport Briefcase:

> *The attitude of a backpack, the form and function of a classic attaché case. Two zip-opening pockets for books, paperwork and legal pads. Snap-closing back pocket for newspaper, magazines and files. Built-in slots for pens, pencils and small tools. Computer disc pocket. Expanding pouches for a calculator, tape recorder or small phone. Comes with our ergonomically designed Comfort-Carry Strap and carries easily over your shoulder. YKK coil zippers used throughout. Entire bag made from a gutsy nylon fabric called Propex. Shrugs off abrasion, punctures and scuffs.*

Nary an adjective in the entire passage. And notice how the writer combines *features* ("slots for pens, pencils and small tools"; "expanding pouches"; "ergonomically designed strap") with *benefits* ("carries easily over shoulder"; "shrugs off abrasion, punctures and scuffs"). It's an approach that allows readers to picture the bag, picture their pencils and tape recorder snugly stowed away in it, and picture the bag over their shoulders. Once they do that, the sale is all but completed.

Use Words with Power

Mark Twain once said that using the right word can be the difference between lightning and a lightning bug. Choose your words carefully, both in your headline and in your letter copy. If you leaf through successful direct-mail pieces, you'll repeatedly see the following words:

Free

New

Easy

Love

Safety

Proven

Breakthrough

Save

Money

You

Guaranteed

Two-for-one

Congratulations

Act now

You should avoid some words and phrases as well, says copywriter Debra Jason. She doesn't like to use the word "will," for instance, as in "You will lose 10 pounds." "I just take it out and write 'Lose 10 pounds!'" she says. "It's more direct and assertive." Another word to avoid is "learn"—it implies work for most people, says Jason. Try words such as "discover" or the phrase "how to."

Keeping your writing as fresh and new as possible, while reassuring people with familiar terms, can be a tricky balancing act, of course. "I try to avoid words that have been beaten into the ground," says Chuck Whitmore, executive vice president for Oxford Communications, a marketing firm that has put together direct-mail campaigns for Dow Jones, Merrill Lynch, and other firms. "Words like 'value' and 'quality' really don't say much to people anymore."

Include Testimonials

Nothing is more convincing to a prospect than the words of someone else who has used the product or service in question. And even though it's true that, deep down, people know that you have carefully picked and edited them, testimonials in a direct-mail piece still add impact. One way to get testimonials is to give people space to write their comments in a question-naire about your product. (This is also a good way to test prospects for more products.) Be certain to get permission, of course; otherwise, you're guilty of misappropriation of a name for commercial purposes, and you might find yourself in court. If you can't get permission, then paraphrase the word-ing and fold it into the copy without attribution; even then the words are apt to have that hard ring of experience that you desire.

Try to use testimonials that are concrete. Instead of "I loved your product!" use "Your wonderful mug kept my coffee hot for more than an hour." If you can, include a photo of the person whose testimonial you're using. It will help convince the reader that the person actually exists.

Keep It Friendly

One of the most difficult tricks in writing is finding the correct tone, what writers call "voice." In direct mail, most professional writers strive for a casual, even informal tone, one that seems human and approachable. Imag-ine having a conversation with the prospect. If you could capture that conversational tone and put it on paper, that's the effect you want. In fact, it's not a bad idea to speak your pitch into a tape recorder or use voice-recognition software, and then transcribe and edit the result.

Of course, finding the right tone is trickier than that. People read words in a different way than they hear them, so words that may sound natural can seem corny or contrived in print. You don't want to go over-board on the casual tone ("By golly! This is the handiest little gadget you'll ever come across!"). To help ensure that you have the right tone, here are some guidelines:

- Use contractions—"don't delay" for "do not delay"; "you can't lose!" for "you cannot lose!" Contractions are chatty little parts of speech; they'll make your copy friendly and approachable.

- Use personal pronouns—"you," "your," and "you're"—and lots of them. You want readers to feel they are being addressed directly.

- Use short sentences. Even fragments. This approach creates a piece that has a faster-paced, more relaxed tone. But don't. Make every sentence. The same length. Because that can. Become quite repetitive. Mix in sentences of medium length to create a more pleasant and natural rhythm.

- Use active voice. Don't write "You'll find our product to your liking." That sentence is in the passive voice and backs into the subject.

Write "You'll like our product!" Here are some examples of passive voice and their active opposites.

Passive	Active
Mailing costs were cut by our new lightweight envelopes.	Our lightweight envelopes cut mailing costs.
The 6 pockets in these pants will let you store essentials.	Store essentials in these pants' 6 pockets.
Zip! Software will allow your child to learn more.	Your child will learn more with Zip! Software.

Put on the Polish

It has been said that good writing does not come from writing; it comes from revising. Don't be happy with your first draft—remember, it exists chiefly to fill space in your wastebasket. Take the time to revise your copy carefully.

First, Walk Away

The next time you write a letter or an e-mail, leave it for 30 minutes while you do something else—walk the beagle, have a cup of coffee, buy stamps. When you return to it, you will almost certainly find a typographical error you missed earlier or a sentence that you now can make more precise and meaningful. You need distance from a piece of writing to really "see" it. And the more distance, the better. Ideally, let your just-finished direct-mail letter percolate in the word processor for at least a day while you deal with other tasks. When you return to it, you'll be able to see it more critically.

Read It Aloud

Remember, you want a friendly, conversational tone, so read the piece aloud. Doing will give you a good sense of its pacing and rhythm. You'll find gaps and errors you might miss when you do not read aloud. Even better, have someone read it *to* you. That way you'll really get a sense of how customers will "hear" the pitch when they read it for the first time.

Trim, Trim, Trim

It's true that no formula dictates that a direct-mail piece should be one, two, or four pages long. By the same token, you don't want to take up any more of your prospect's time than is absolutely necessary. Use the word count feature in Microsoft Word 2000 to determine the length of your piece, and cut that by 10 percent. Then cut it by another 10 percent. Take out every word that does not enhance the pitch, strengthen the offer, or explain the benefits. Be ruthless—heaven knows your customers will.

Using Word to Improve Your Writing

With Microsoft Word 2000, you won't have to spend precious time worrying about document presentation or layout. Your direct-mail letters, brochure copy, and other materials can be easily created in Word 2000 and sent directly to a printer for publication. Improved intelligent features, such as background spelling and grammar checking, will be particularly useful in highlighting errors that might get by even the best copywriters and proofreaders. And you'll appreciate the ability to create and use various templates, which allows you to create or update letters and other documents more easily.

Word 2000's Track Changes feature also helps you "pass around" a document for review and revision. Better integration with Outlook enables a document to be easily e-mailed to coworkers. Easy-to-create styles allow you to develop formats for different direct-mail pieces without having to reset fonts and type sizes in the documents. Just select the text you want to change, click the Styles drop-down arrow on the Formatting toolbar, and then click a preset format.

Finally, Word 2000's powerful formatting capabilities allow you to fine-tune your copy's appearance, add spacing or indent variety to letters, insert headlines, and add spot color for real impact.

New Features in Word 2000

Word 2000 is designed to build on the ease-of-use and power of Word 97 while adding new, useful features. In particular, Word 2000 is designed to intelligently adapt to the way you work, so you can stay focused on the task at hand rather than making minor corrections or finding and correcting problems. Among the features that you'll find useful while writing your direct-mail copy are these:

Click and Type. Double-click in a document to insert text where you want it. You'll see what the formatting will look like through a feature called cursor hinting, and the correct formatting will be applied automatically (for instance, text wrapping and tabs).

Word 2000 also makes it easier to format your document yourself, such as centering headlines in a letter, indenting paragraphs to help break up copy, or creating a single line of text with the text in different areas, such as left-aligned text on the left and right-aligned text on the right.

WYSIWYG (what you see is what you get) Font list. The drop-down Font list on the toolbar shows you what a particular font looks like before you apply it to your document. You save time finding a font that expresses perfectly the look and feel that you want your direct-mail letter to convey.

Picture bullets. Bullets can be used to set off and highlight short lists of features in a direct-mail letter or brochure. Word 2000 includes a number of useful bullets, including picture bullets. If you find graphical elements that you like—such as bullets on a Web page—you can copy and paste them in your own document. Once copied, the picture bullets are just like regular Word bullets, conveniently repeating automatically in lists.

Collect and paste. Word 2000 makes it much easier to collect text from several locations and drop it into a single document. Now, when you copy a text selection, a graphical Office Clipboard appears. The Office Clipboard can hold up to 12 items; just place the insertion point at the location where you want to place the copied text, and then click the Clipboard icon for the item you want to insert.

Quick file switching. Word 2000 also provides faster access to individual Word documents by placing a button for each document on the Windows taskbar. This helps you move easily between documents, particularly during collect and paste operations.

Self-repairing. New, more intelligent Microsoft Installer technology makes Word 2000 capable of fixing itself. When you start Word, it will determine if essential files are missing and where they can be found; then it reinstalls the missing files with little or no headaches for you. Word 2000 also can repair corrupted font files or missing templates by scanning Office files for discrepancies and in many cases fixing the problems.

In short, Word 2000 helps you give your direct-mail customers "the word" on your products and services.

Chapter 5

Producing the Direct-Mail Package

Not long ago, John White was asked by Time Warner to produce a direct-mail package promoting the company's cable channels to prospective advertisers. Some executives had cooked up a tentative design for the package—a slick folder with pockets for various printed materials. "It was a 'me-too' package," says White, founder of Perceive, an award-winning Long Beach, California–based design and advertising firm that frequently puts together direct-mail packages. "It looked just like what everyone else does. I went back to them and said, 'Look, you don't want to do that. You want to take all this stuff and organize it so that anybody can use it.'"

White's solution was to create a cardboard wheel that could be turned to open different windows. As the recipient spun the wheel, in 20 different windows appeared a different Time Warner channel—CNN, for instance—with the audience demographics, secondary audience, secondary cable channels, and other information of use to media buyers. "It was completely user friendly, and it got an incredible reaction from the client," White says. "It sold itself—the Time Warner salespeople could come in and their work was half done."

White's solution may be a bit exotic for the entrepreneur just trying out direct mail, but the principle he applied can be used for any mailing: Recipients want something that's useful, is easy to understand, and meets their needs. Whether you're selling a product or service, generating leads, or disseminating company information, a direct-mail package needs to meet those criteria.

That's where good design plays a key role. It is true that getting your direct-mail piece into the right hands (by using the right list) and making it tempting to buy (with the right offer) are the most important parts of the equation; however, design can enhance the effectiveness of your pitch. In this chapter you'll learn about some key design (and more copy) considerations for the different parts of a "classic" direct-mail package: envelope, letter, brochure, and response card. I'll also touch on postcards, self-mailers, and catalogs; discuss how to get the best outside help; and show how to use Microsoft Direct Mail Manager to help get your mailings delivered.

Start a Cheat Box

Sure, originality is great. But remember two principles: (1) the wheel has been invented and (2) no subsequent improvement has done away with the need for roundness. The point is that you can *learn* from what other people have done, adding your own incremental improvements without starting from scratch. Start a "cheat box"—a cardboard or plastic box into which you start tossing all your jun … er, direct mail. Sad to say, perhaps, but you'll probably have a sizable pile within a week. Within a month … well, enough said. Start looking through it. Soon you'll see certain visual and typographic patterns. You'll find that some direct-mail pieces will draw you in, persuade you, and convince you to buy. Others won't. Borrow the elements you like, and avoid the ones you don't.

The Envelope

Think of the envelope as the doorway to your direct-mail piece. It's the first thing your prospective customer will see, and you want to make a good impression. Think about the sort of approach you want to take. Your choices could include the following:

> **Be mysterious.** Imagine a blank white door in a wall. What could be behind it? Curiosity will lead many people to open it. And curi-

osity will lead people to open a white envelope with no return address. Sure, they know it's probably a marketing pitch, but what if it isn't? What if it's from that cute girl you knew in high school, and you always wondered what happened to her, and she *finally tracked you down!* Oh, it isn't from her. But look! Here's a big discount for window screens! And my house could use some new ones. Bingo.

Be provocative. Another door might have a window in it. You look inside, and you see that people are having fun! Of course, you want to join them. Write envelope copy that leads the reader to believe the envelope's contents will be enjoyable. Don't simply write "Big savings inside." Everyone has seen that, and everyone has been disappointed. Use big, eye-catching headlines, and make a promise that opening the envelope will be rewarding for the recipient. "Savings of $100 inside!" (restaurant discount coupons) or "Inside this envelope is a new way to view the world!" (binoculars for a sports or wildlife fan) or "David is inside, and he wants to have a word with you" (trip to Italy, including a visit to Michelangelo's famous statue in Florence).

Be colorful. Perhaps your door is washed in warm, inviting colors. Your envelope too could be warm and colorful. Color will help an envelope stand out from most mail, and it may even evoke pleasant feelings from the recipient. What could be more inviting than an envelope splashed with the colors of spring, mailed out in the middle of a cold, wet Seattle winter? Figure 5-1 shows the envelope for a mailing sent out by Schwartz Brothers Restaurants in Seattle. Its lavender, orange, and yellow color scheme was the perfect antidote for the wettest winter in 50 years.

Be different. Doors are always upright rectangles. Nobody, though, said that all envelopes have to be No. 10. In fact, postal regulations allow a fair amount of leeway in envelope dimensions. Make your envelope square, or triangular, or larger than usual. Again, the point is to make it stand out. The only caveat is that custom envelopes can add considerable cost to a mailing; your print shop should be able to help you find a standard size that's still a little off the beaten No. 10 path.

Be honest. When we're wandering through a building, we appreciate it when doors are clearly marked. Perhaps you can take the direct approach with your envelope. "OK, what you're holding

Figure 5-1

This envelope for a direct-mail insert sent out by Schwartz Brothers Restaurants used an unusual shape and was washed with warm colors to be both eye-catching and inviting.

is a piece of junk mail. But we think this one is different. Open it and find out why." People are apt to be disarmed by such an approach, and they may open your envelope just to see what the heck could lead you to make such a startling statement.

The Letter

In Chapter 4 we discussed tips on writing copy for your letter. Design is important, too. Its goal is not just to make your letter look spiffy, but to ensure that you make your offer clearly, as many times as possible, and that you encourage readers to buy. Be sure to include the following in your letter:

A Johnson box. Who knows where this term came from, but there it is. The Johnson box is a block of copy that appears at the top of page one, above the salutation. It should be short and punchy and should serve as a teaser for what's to come in the letter, introduce the offer, or highlight a feature or benefit. Here's a Johnson box written by The Hacker Group for AT&T Wireless Services:

> *If you think high-quality cellular phones and service are too expensive—call 1 800 585-1658 now to find out about our new AT&T Wireless Services package!*

That copy appeared in an italicized font and red ink—it was the most eye-catching element on the page (see Figure 5-2).

AT&T Wireless Services

If you think high quality cellular phones and service are too expensive – call 1 800 585-1658 now to find out about our new AT&T Wireless Services package!

Figure 5-2

A Johnson Box is a copy block at the top of the letter that announces and reinforces the content of the letter.

Lots of white space. People won't read a single-page letter that's dense with copy, but they will read a two- or three-page letter. Be sure to format your copy so that it has frequent paragraph breaks that add white space.

Add clutter. A little clutter, creatively applied, adds visual interest and makes it look like something *is really happening!* Good clutter can include small headlines scattered throughout the copy, important lines that are underscored for emphasis or **written in boldface** or **both when you really want to make a point,**

or testimonial quotes scattered along the margin. Remember, be judicious. Too much clutter becomes, well, clutter.

Add art. Be sure to include a small photo or image of the product or a depiction of the service you provide. Art adds visual interest.

P.S. As noted in Chapter 4, a postscript is an important element in a letter. In fact, it may be all the recipient reads. Be sure to add a P.S., and in it repeat your offer, the key features or benefits of your product, and what the recipient must do to take advantage of your generosity.

Typographic Matters

The personal computer makes anyone a typesetter. That's both a wonderful thing and a terrible thing. The hundreds of possible type styles now available can create a letter that's like a visual Frankenstein—bits and pieces of things that simply don't belong together. And it's liable to send your customer away, screaming in frustration, if not downright terror. Here are a few rules about typography.

Mind your serifs

Typefaces are families of a particular type style (for example, Franklin Gothic or Times New Roman) that come in one of two varieties: serif and sans serif. You're reading a serif typeface now, one with the fine cross lines at the bottom and top of a letter, as you see in this capital R. Sans serif typefaces, such as Franklin Gothic, don't have those cross lines, as you see in this R.

Generally speaking, sans serif typefaces make good headlines, while serif typefaces are best for body copy. In fact, reading blocks of sans serif type is much more difficult than reading the same words printed in serif type.

Don't mix fonts

Given the wealth of type fonts, which are the different variations of a typeface, it's tempting to start mixing them for visual variety. You might be tempted to use both Franklin Gothic Medium and Franklin Gothic Condensed. Don't. Limit your letters to three fonts maximum. Two is better. Use one for your Johnson box, headlines, and perhaps testimonials, and another for the body copy.

Size matters

You may be able to economize by using a slightly smaller type size, say 9 points instead of 10 points. Don't. As a rule, 10-point type is the smallest you should use for body copy. If you know your material will be read by people over 40, you may want to run it in 11- or 12-point type. Lots of people have difficulty reading small print, and by the time they find their reading glasses, they may say "the heck with it."

Layout Tips

- Avoid clutter. Keep pages clean and neat, with ample white space.
- Avoid fancy typefaces. If you are using multiple typefaces, it's best to set headlines in a sans serif type, with body copy in a serif type. Mixing sans serif and serif typestyles can be distracting.
- Use white space and subheadings to break up large pieces of text.
- Narrow columns of type look longer than the same amount of type in a wide column.
- Use typographical devices such as dashed lines to indicate where a coupon should be cut.
- Design for your audience. Young people like edgy, somewhat busy layouts. Older readers prefer a more formal look.
- Be sure the "look" of your various pieces—envelope, letter, brochure, and response card—is complementary. Don't change typefaces or logos between pieces.
- Don't forget to sell your product. It's easy to get so caught up in producing great "creative" work that you lose sight of making the sale.

The Brochure

Many direct-mail packages add a brochure to the mix. In general, the letter makes an emotional, personal appeal to a prospect, while the brochure provides rational, technical information about the product. In other words, the letter *sells,* and the brochure *tells.* Let's say you are marketing a new style of lawn mower. In your letter, you'll tell prospects how much easier this mower will make their lives (remember: sell the hole, not the drill), how they'll have more time to spend doing what they really want to do, and how little it will cost them ("just four easy $99 payments!"); you'll also include any other offer strengtheners ("satisfaction guaranteed"). The brochure can speak to the part of a person that wants to know all the cool technical stuff about the mower. How many horsepower? How wide a blade swath? What kind of ignition? The readers may not understand half of it, but it appeals to their desire to know quite a bit about this mower's features before committing nearly $400 to it.

When do you need a brochure? Whenever your product or service is more visually and technically complex than you can convey clearly in a letter. If you run an accounting or copywriting service, for instance, you probably don't need a brochure. You may want a a brochure if your product or service has any of the following traits:

> **It is visually complex, or it consists of several parts.** For our hypothetical lawn mower, for instance, a brochure could contain a photo of the lawn mower, another of its easy-to-use ignition, still another of the underside and the fearsome-looking blade.

> **It can be used for different things.** Sure, your mower mows. It also mulches, can get into small spaces, and is powerful enough to cut even heavy weeds! Show that in a brochure.

> **It is a "considered" purchase.** Whatever price point or product makes readers ask themselves "Do I really need this?" is a considered purchase. It may even be a *grudge* purchase— something somebody may not want to buy at all. Brochure information that convinces people that the product or service offers value, is worth the money, and can improve their lives will help overcome these barriers. In short, the brochure should tell recipients what they need to know to overcome their reluctance to make the purchase.

> **It looks great.** We're a design-conscious society. A product that looks powerful and efficient will outsell one that doesn't. If your mower looks like a wonderful, high-tech lawn tool, a brochure can show it to advantage.

Keep in mind, though, that you don't want to ask a brochure to do too much. Focus on a single product. If you offer an array of products—a weed-cutter, a siding-washer, and a rototiller in addition to a lawn mower—you should produce either separate brochures for each (with cross-references) or a small catalog.

Brochure Basics

The brochure is one of the most visually complex pieces in a direct-mail package. (The word "brochure" is French, by the way, from a word meaning "to stitch.") You may find it useful to hire a designer for this part of the project. Plus, once a designer has done your brochure, visually tying the brochure to your letter, envelope, and response card will be simple.

Publisher 2000 and the Brochure Wizards

Still, the do-it-yourselfer can publish a good brochure. Microsoft Publisher can be a powerful tool for designing effective and affordable brochures. Publisher has scores of brochure wizards (see Figure 5-3) that can be fine-tuned by changing the color schemes or modifying the typefaces.

Publisher's brochure wizards include preset locations for inserting photos or illustrations, as well as text boxes that offer suggestions on the sort of copy to use in that particular box. It's easy either to print the brochure yourself or to export the design to a floppy disk or Zip disk to take to your local print shop.

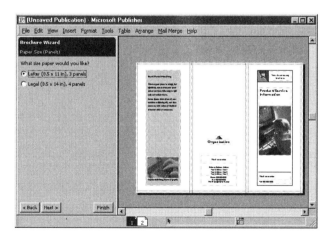

Figure 5-3

A sample brochure wizard.

Useful tools for a project such as making a brochure (or a catalog; more on that shortly) include the following:

- A good-quality scanner and 35mm camera with a short telephoto lens (90mm with macro capability is ideal) or a short zoom lens that offers that same focal length.

- Alternatively, a digital camera capable of producing high-resolution images.

- Photo-editing software such as Microsoft PhotoDraw.

- Clip art catalogs. Microsoft Publisher includes an excellent clip-art catalog with hundreds of images in a variety of styles. Broderbund's ClickArt clip-art series includes packages with as many as 300,000 images.

The Cover

The first thing a reader sees will be the brochure's cover. For a direct-mail brochure, the cover should clearly illustrate the product and in some fashion state both the offer and the product's key benefit. Use a headline that will lure the reader to go further.

The Inside Spread

Once opened, the brochure should highlight three or four important features and benefits of the product. Photos of our super mower, for instance, may show it plowing through tall grass, maneuvering in a tight space, and folding for easy storage in the garage. A single headline across all the inside panels will help tie the different visual elements together. (This is an old trick—the Dutch Master painters always had their group portrait subjects stand and sit around a long table that provided the visual organization.) Different colors may help set off each panel.

The Back Page

The back page is the place for real hard-core data: the exact specifications for your product. Also, here's where you want to restate your offer, include a toll-free number, and mention payment options.

The Response Card

The last piece in the direct-mail package is the response card or order form. Don't neglect this piece. After all, it's the vehicle a customer will use to give you money. It's perpetually astonishing to see how many companies make it difficult for customers to empty their wallets on the company's behalf: too few check stands, slow lines, limited payment options, you name it. Listen, giving you money *should be the easy part* for the customer. Also, keep in mind that the response card may be all that a recipient reads; some people skip the letter and brochure and cut right to the chase: What does this guy want me to do, and how do I do it?

The response card is also a powerful data-gathering tool for you, so think carefully about what goes on it. Your response card should contain the following:

The offer. Restate your offer here, whether it's dollars off, two-for-one, or any other special offer. And give the customer a chance to respond affirmatively: "YES! I want to buy your amazing new Handi-Mower. Please bill me in four installments of $99."

Customer reassurance. The customer doesn't know you but is being asked to send money. Reassure the customer with a guarantee or a money-back promise, prominently displayed.

Involvement devices. In some cases people seem to *like* doing a little work that, strictly speaking, is unnecessary. One common example in direct mail is a peel-off sticker that the customer affixes to the response card.

Room for long names. "Smith" is a common, short name. But plenty of names are longer, and remarkably few response cards allow enough room for those blessed with names like "Gantenbein." Also, if you have hired a firm to compile your database, ask its representatives about entry fields and its preferred format for customer responses.

A chance to say "maybe." Don't make the response card a strictly "yes/no" proposition. Some people who are interested in your product or service may not be prepared to buy at that moment. Let them check a box that says something like "No,

I won't be buying at this time." If they take the time to check the box and send in the card, they are showing considerable interest. You can also use this information to determine how to improve your *conversion rate*—the rate at which people make a purchase.

Self-Mailers, Postcards, and Catalogs

Direct mail doesn't end with the "classic" package. As long as the U.S. Postal Service allows it, you can mail almost anything. Some other common tools of the direct-mail marketer are self-mailers, postcards, and catalogs.

Self-Mailers

For a direct-mail campaign touting its high-end Model 88 table radio, Massachusetts-based Cambridge Soundworks opted to use a *self-mailer.* This is simply a mailing piece that is not inserted into an envelope. The address of the recipient is stamped or fixed directly on the piece.

Self-mailers can be less expensive to produce than a classic package. The Cambridge Soundworks self-mailer, for instance, was an all-in-one piece, combining envelope, letter, brochure, and response card. Still, it was an elaborate piece: a full-color mailing printed on quality stock that cost several dollars per copy to print (its back page is shown in Figure 5-4). You need not be that elaborate; in some cases an $8\frac{1}{2}$-by-11-inch sheet of paper, folded two or three times, can be used as a self-mailer.

Self-mailers, however, have drawbacks. Foremost among them is that they immediately identify themselves as advertising. (Do any of your friends, for instance, send you self-mailers?) You have to work hard to pass the reading-mail-over-the-trashcan test. Cambridge Soundworks worked to overcome that by relying largely on its house list—people who were familiar with the company and had even purchased its products. A mail recipient who has never heard of you may reflexively toss the piece. In fact, Seattle direct-mail expert Bob Hacker says that classic packages often out-respond self-mailers by 100 percent or more.

Figure 5-4

This self-mailer for Cambridge Soundworks has an eye-catching design and considerable information about the product—a high-end table radio.

Postcards

Postcards identify themselves as advertising even more quickly. Moreover, they're easily lost amid the mass of mail that comes into most homes or businesses each day. Still, the postcard can be produced incredibly cheaply and quickly, and for tasks such as a simple announcement ("Half off all our stock!") or directing recipients to a trade-show booth, they're useful tools.

Moreover, this cheap and quick method need not be dull. For instance, when high-tech startup Paraform, maker of sophisticated 3-D modeling software, wanted to draw people to a trade-show booth, it assembled in a matter of days a simple two-color postcard with a quote from Leonardo da Vinci that John Doffing, director of marketing, felt tapped into Paraform's corporate ethos. "Simplicity," the postcard cover read, "is the ultimate sophistication." (See Figure 5-5.)

Mailed to names mined from the company's Web site, the "Simplicity" postcard got a huge response. Traffic to the booth was heavy, and company salespeople often see the postcard tacked to customers' office walls. "People are asking for posters of it," says Doffing.

Figure 5-5
Postcards can be cheap and quick direct-mail solutions.

Catalogs

Needless to say, catalogs are the antithesis of postcards—they are potentially complex, expensive, sophisticated productions that entail serious commitment on the part of the direct-mail marketer. "Cataloging is hard," says Dan McIntyre, a catalog consultant based in Portland, Oregon. "Cataloging is expensive. Cataloging is a costly way to do sales. It may not have been that way 30 years ago, but it is now." Competition for catalog customers' attention is fierce, as more major retail players seek to leverage their brick-and-mortar stores through catalogs. By some estimates, more than 14 *billion* catalogs are mailed out each year. Even seasoned catalog veterans such as Lands' End have struggled in the recent catalog-selling environment. And costs for staples such as paper and catalog mailing lists continue to rise.

Moreover, says McIntyre, who with his wife, Susan, runs McIntyre Direct, the cost of catalogs is such that it's all but impossible to make money on single sales. "You lose money when you're growing," he says. "You begin to make money only when you've converted names from rented lists into a house list and made those people repeat customers." That can make life difficult—it means that you have to finance the catalog startup through some avenue other than catalog income, as it may be some time before a catalog becomes self-sufficient.

Still, of the 10,000-plus different catalogs offered in the United Stat perhaps as many as 90 percent are published by entrepreneurs. And cat: log shoppers tend to be avid readers of catalogs and fairly regular buyers. If you can get someone to purchase from a catalog, and if that customer is satisfied with the product, chances are that he or she will become a repeat customer. A particular challenge for beginning catalog mailers is the sheer volume of catalogs they must compete with, as many companies that formerly relied on brick-and-mortar retailing enter the catalog fray.

When to try a catalog

The great advantage of a catalog is its ability to offer a large number of relatively low-priced items to a customer. As a general rule of thumb, in fact, catalogs should be used when you have quite a few products to offer. That isn't a hard and fast rule, of course—a catalog could be designed around a single, high-ticket item and reinforced with pages of accessories and options. Nor should you plan on producing a catalog of items that cost less than $10 each unless you are prepared to watch costs with a microscope and perform all of the manufacturing and order fulfillment yourself.

To some extent, you are constrained by the laws of publishing. Catalogs can be printed in eight-page batches. But, warns McIntyre, anything smaller than 16 pages is apt to do poorly simply because it is too hard to match prospective customers with such a limited product base. "Even if you get the catalog in exactly the right hands," he says, "that customer is apt to say, 'I don't like this silk blouse, what I need is a linen blouse—but they don't have one.'" Sixteen pages is the safe minimum for a startup catalog, he says. Twenty-four pages is even better.

Catalog Basics

Whether or not you design your own catalog, says McIntyre, you should keep several principles in mind. It isn't enough to simply find some catalogs you like and try to copy them. Too often, people will miss the entire point of what the "model" catalog is doing.

One cardinal rule is to take advantage of "hot spots" in a catalog. Hot spots are where your strongest products should go, as these are the pages most apt to be seen with even a quick thumb. The cover is the hottest spot, of course, followed by the back page. Pages two and three—the inside cover spread—are hot spots, as are the equivalent pages just inside the back page (a surprising number of people read catalogs back to front). Also, the center spread where the staples tend to force a catalog open are "hot."

You also need to pay close attention to how a customer's eye will tend to flow across each two-page spread. A number of studies on eye flow show that eyes tend to move along the path of a large "V," placed on its side with the point of the "V" at the far left side of the spread. "Any product along that line will sell better than products outside of it," says McIntire. "With a well-designed catalog, you can blur your eyes up and look at the page and see that 'V'—designers design for it. If you look at an amateur catalog, everything lines up down the center. That stops the eye, and you won't sell a thing outside that line." Similarly, the upper-left and lower-left corners of a two-page spread tend to be "death valley"—best for products that are not your strongest.

One last point is to understand that a cataloger's best intentions can work to subvert its customer's ability to absorb information. "People spend six months producing a catalog, and they want people to spend six months reading it," says McIntyre. "But people scan—they don't read every word." So, says McIntyre, while you want to play off a catalog's ability to tell a strong story, you also want to be sure that a reader can pick up key points simply by scanning headlines and subheadings. "You need to tell your story layer by layer," he says.

A Day at Lands' End

Lands' End, based in Dodgeville, Wisconsin, is recognized as one of the nation's leading catalog companies. Some facts about its operation:

Toll-free phone lines to sales and customer service are open 24 hours a day, 364 days a year. The company closes only for Christmas.

On a typical day, customer-service representatives on nearly 300 phone lines handle between 40,000 and 50,000 calls.

During the weeks prior to Christmas, over 1100 phone lines handle more than 100,000 calls daily.

Lands' End receives 14 million calls annually.

Sales representatives receive 70 to 80 hours of product, customer-service, and computer training when they are hired and 24 hours of training each year thereafter.

In-stock items leave Lands' End's Dodgeville distribution center (the size of 16 football fields combined) the day after they're ordered.

Designing a Catalog

Microsoft Publisher 2000 is a useful tool for developing a catalog. It comes with a variety of catalog wizards that provide an excellent starting point for conceiving a catalog.

You can save money on the launch of a catalog in numerous ways:

Write your own copy. Look through your cheat box and find a catalog that is similar in style to what you would like yours to be. Now put a sample of your product on your desk and take a shot at writing copy that sounds like the example you like. It may take a few drafts, but chances are you can come pretty close to your desired result. You may still want to hire a copywriter, but do so only to have him or her edit and improve what you've already produced.

Take your own photos. Today's cameras offer loads of sophisticated features that make it easier than ever to take good-quality photos. Use a 35mm single-lens reflex camera (not a point-and-shoot camera), purchase a powerful flash that mounts atop the camera (preferably with an extension so that the flash is as far from the lens as possible; this makes for more pleasing lighting), and use a short telephoto lens. And follow these three cardinal rules of photography:

Get close.

Avoid busy backgrounds.

Use the "rule of thirds," positioning key photo elements in locations where lines would intersect if three horizontal and three vertical lines divided your picture frame.

Hire on the cheap. You can find the stars of tomorrow at your local college or university today. Advertise for photographers or designers in the campus newspaper, or post a notice on the advertising department's bulletin board. You'll pay a fraction of what you would for an established "professional."

Use stock art. Many clip-art CDs now come with high-quality photographs that can enhance a do-it-yourself catalog.

Be creative. No, you can't afford to fly to the Galapagos for a photo shoot. But you can still be original, says Oregon-based catalog expert Susan McIntyre. Find ways to add value to a catalog by including useful information, testimonials, and inexpensive graphic elements such as colored backgrounds and type.

Creating a catalog for your small business can be a challenging task. But with patience and the right product mix, a catalog can be a successful marketing formula.

Chapter 6

Printing and Mailing

K/P Printing, not far from downtown Seattle, is redolent with the smell of ink and resounds to the clatter of the big Heidelberg sheet-fed presses. K/P prints for clients ranging from Microsoft Corporation to Hewlett-Packard, and it is a direct-mail specialist. In fact, K/P is something of a rarity: It can take a direct-mail marketer's finished design and print it, address it, mail it, and handle order fulfillment. First and foremost, though, K/P is a print shop. The process for getting your direct-mail piece printed at K/P is much like the process used at print shops across the country. Let's take a closer look at the process K/P uses as an example of what a good print shop can do for you.

How Much Time Is Needed?

You want to allow plenty of time to complete your direct-mail piece—a month if you have a finished camera-ready design, says Bob Hetzel, K/P's direct-mail manager; two months if you're starting from scratch. Haste creates all sorts of problems. You're apt to miss errors in the proofs, and worse, you may miss your hoped-for mailing deadline and be forced to use first-class mail.

A typical schedule might look like this. Note that some tasks can be performed simultaneously, such as ordering mailing lists while working on the writing and design for your package:

Developing sales and marketing concept	1 week
Copywriting and rough layouts	2 weeks
Finished layout	1 week
Hiring photography or acquiring stock art	2–4 weeks
Print preparation (developing mock-ups; proofs)	1–2 weeks
Ordering mailing lists	2 weeks
Mailing services	1 week
Time until postal service delivers package	1–2 weeks

Preparing Your List

The first step is to clean up your lists and prepare the names on them for transfer to the mailings. In many cases, you may work with a computer service bureau for this task. Its job is to work with your lists to ready them for mailing and build your house and response lists. You can find service bureaus in the yellow pages of your telephone book, or you can contact the local chapter of the Direct Marketing Association for information about local vendors. Your list broker may also be able to recommend a service bureau. It's useful to get references for your bureau—find the names of some companies that use it, and double-check the bureau's work. You are trusting the service bureau with your data, and in direct mail data is vitally important.

Bob Hetzel says a service bureau will typically perform these tasks:

- Standardize lists that come in different forms to ensure that all the names and addresses are in the same format. They then can be folded into a single database.

- Eliminate duplicate names and addresses. It's a pretty safe bet that if you're using several lists, many names will be duplicates. You don't want to waste postage—and annoy a potential customer— by sending multiple mailings. A service bureau will perform a painful-sounding procedure called a "merge/purge." This step compares names across several lists, eliminating duplicates so that no names appear more than once.

- Verify addresses. Addresses can get goofed up. A service bureau will run your list through software developed by the U.S. Postal

Service called the Certified Accuracy Support System, which ensures that the addresses on the list are indeed deliverable.

Making a Mock-Up

The second step in getting your direct-mail piece into customers' hands is to complete what's called a "mock-up" of your mailing. Print a sample of the mailing you've created in Publisher 2000, or ask your graphic designer to complete a mock-up for you. Use the paper type you think you want to use; one role of the mock-up is to estimate the weight of the mailing. If it's a light mailing, going over the one-ounce line can end up costing you a lot of money, says Hetzel.

The mock-up also gives you a chance to double-check your specifications—the dimensions of the piece, how it is printed, and other factors. Make sure it's designed properly for machine-stuffing; a mailing that fits just fine into a No. 10 envelope when you are fiddling with it may be just a touch too big to fit into an envelope via machine. And that can create a big headache for a 10,000-piece mailing.

Choosing Paper

A really successful direct-mail piece hits on all cylinders: The copy resonates with the recipient, the visuals are attractive, and the "feel" of the mailer appeals to the reader's sense of quality. Think of that feel as the direct-mail equivalent of the solid thunk of a well-made car door.

Paper comes in a bewildering array of weights, styles, and coatings. Here again, your cheat box will help. Find some mailers with paper you like and haul them to your printer or designer for recommendations on similar paper styles. Even then, you're apt to confront several choices within the same paper type. Here are the basics about paper.

Weight and Thickness

The rule of thumb is that the heavier the paper, the greater the feel of quality. Weight is given in pounds and is calculated on the basis of 500 sheets of the paper's *standard* size. Because standard sizes vary, the somewhat confusing result is that a cover stock may be rated as lighter than a text stock it actually outweighs. To make matters more complicated, the size of the paper you actually purchase may not be its *standard* size.

As a common-sense rule of thumb, thicker papers weigh more, says K/P salesperson Alicia Miller. But some fairly thick papers are less dense—they have more "air"—than thinner papers. This may be useful to know because weight will be the prime determinant of how much you have to fork over to the postal service to mail your piece. You might be able to use a thick, nicely textured paper that weighs less than slightly thinner paper.

Coatings and Brightness

Paper is either coated or uncoated. Coated paper is the glossy or semi-glossy stuff; *National Geographic* is printed on glossy paper. Why? Because it reproduces photos and artwork better. Uncoated paper is what you find in most books; it's cheaper and perfectly fine for text. In fact, it's easier to read text on uncoated paper because there's little glare. In the direct-mail world, a common combination is to use uncoated paper for the envelope, letter, and response card and coated paper for the more visually elaborate brochure. If you're reproducing color photos or color illustrations, you'll want coated paper.

Paper also comes in different brightness levels, from 1 (the brightest) down to 3. The brighter the paper, the better the reproduction. In most cases, you'll want to use paper with No. 2 brightness for your direct-mail piece. A No. 3 brightness might be fine for some bulk mailings, and No. 1 could be useful for highly targeted mailings in which you want to make a particularly good impression, says Miller.

Glossary of Common Printing Terms

Basis size. The size of a full sheet of paper in a given paper family. *Cover* basis size is 20 by 26 inches. *Book* basis size is 25 by 38 inches. *Bond* basis size is 17 by 22 inches.

Basis weight. The weight of 500 sheets of paper in a given paper family. Assuming the paper is of the same family, brochures printed on 40-pound paper will weigh twice as much as brochures printed on 20-pound paper.

Bleed. An image that extends to the edge of a sheet of paper. In most cases, the sheet will need to be trimmed to create the appearance that the image extends beyond the edge.

Brightness. A measure of the reflectivity or whiteness of a sheet of paper. Brighter paper typically costs more.

CMYK. The four colors used in full-color printing: cyan (blue), magenta (red), yellow, and black.

Color separation. The process of converting a full-color image into the four basic printing colors.

Comp. A rough color proof.

Contract proof. A color proof used for final approval.

Crop marks. Small lines at the corner of page that indicate where it should be trimmed.

Embossing. Pressing a relief image onto a paper sheet to create a raised image.

Font. The complete collection of the characters in a single type design—ABC, abc, 123, #$%, and so on.

Four-color printing. Printing with the four printing colors (CMYK), which results in the appearance of *full* color.

Four-up. Printing four identical images on a single sheet of paper, which then is cut into four separate pieces.

Halftone. A black-and-white image that is made to contain shades of gray by varying the size and number of tiny dots of ink. Black-and-white photographs in newspapers are created with a *halftone* process.

Makeready. The time between the start of the press run and the production of the first good sheet. This process can generate a fair amount of wasted paper.

Pantone color. A color-matching system developed by Pantone, Inc. Pantone colors are assigned names or numbers to ensure uniform reproduction.

Postpress. The process of trimming, binding, gluing, and packaging printed products.

Prepress. The part of the process up to the time the presses begin to run.

Register mark. A target repeated in the same location on each color plate. When the targets are perfectly aligned from one plate to the next, the color is said to be in "register."

Spot color. A color other than black, and usually not one of the three primary process colors (cyan, magenta, yellow). Spot color can be used to enhance the appearance of a printed document without the expense of full-color printing.

Choosing Ink and Colors

If it exists in nature, chances are you can find the color in what is called the Pantone Matching System, or PMS. Developed by Pantone, Inc., PMS is a system for rationalizing color selection and ensuring consistent results from printer to printer. Books with Pantone color selections are available in the graphic arts sections of most bookstores. Publisher 2000 also uses the PMS system so that your documents are printer-ready when they leave your computer. Most good printers also can custom-match any color you bring into their shops.

The Final Steps

Once you've completed a mock-up and decided on paper and ink selections, it's time to find out how much the print job will cost. At most print shops an estimator will work with you to determine that figure. After the estimate is complete, the job goes to the press team.

Before the final printing occurs, you'll be asked to review a proof of the printed product, usually called a "blue line" because of the blue tint in the film commonly used to create these proofs. Take the time to go over it carefully—any mistakes you miss now will be corrected at your expense. Ideally, have a coworker double-check the proofs along with you. It's amazing how many mistakes an extra pair of eyes can pick up.

Saving Money on Printing

Printing is not cheap; a mailing that costs only 50 cents per piece to print will result in a $5,000 printing bill if you print 10,000 copies. Pennies add up. There are several ways you can save money on printing, says Howard Oberstein, a print broker and consultant in Arleta, California:

- Shop around. Get bids for your printing from two or more print shops. But don't shop on price alone, Oberstein says—a printer who knows you can be a valuable ally in what is a complex process. "The person who always makes price the number-one consideration doesn't build relationships," says Oberstein. "There are times in the process when you need a friend out there."

- Keep your design as simple as you can without losing impact. For instance, "bleeds"—art images or colors that go right to the edge of the paper—will cost you more, because of the extra paper required to print them, and may add little to the design.

- Submit your design to the printer in electronic form, ready to go into the printer's computer and from there to the press. Talk with the printer beforehand to ensure that your system is compatible with the printer's. "We had a crisis the other day when a print job that was supposed to have a duotone [a two-tone color shade] didn't print as a duotone," says Oberstein. "It hadn't carried from the client's diskette to the printer's computer." Fonts also can be a problem; use the Pack And Go Wizard in Publisher 2000 to ensure that your fonts are copied to a floppy disk or Zip disk with your print job and appear as you intend.

- Try to ensure that your design makes efficient use of paper. A small change in dimensions may allow you to squeeze another brochure or letter out of a sheet of paper.

- Match the printer to the job. If you're printing a fairly simple package with an envelope, letter, and response card, you don't need a big press. Your local quick-print shop may be just fine. Indeed, for small mailings of a few hundred pieces or fewer, you may find that one of today's high-quality inkjet printers is perfectly adequate for printing the job yourself.

What's Your Break-Even Point?

One of the great things about direct mail is that, as with other marketing devices, simple mathematics can tell you how much you can expect to make and how many responses you require to begin making a profit.

To determine your "break-even" point, the point at which you cover your costs and begin to make a profit, you must first calculate the following:

- Current selling price of product

- Variable cost of product (cost of merchandise + shipping expenses + handling expenses + additional expenses)

- Gross margin per unit (selling price – variable costs)

- Fixed expenses (salaries, rent, other costs of doing business)

- Direct-mail expenses (cost of printing package, lists, premiums, postage, processing and filling orders)

Your break-even point, then, is based on this formula:

$$\frac{\text{Total fixed expenses} + \text{direct mail expenses}}{\text{Gross margin per unit (selling price} - \text{variable costs)}}$$

Using Microsoft Excel, it's easy to create a worksheet that shows you how small changes in your selling price or costs can make a big difference in your break-even point. Figure 6-1, for instance, shows one possible scenario. Each unit, a nifty salt and pepper shaker set, sells for $39.95. Variable costs for each unit total $21. Fixed costs and direct-mail costs come to $17.050. With those figures, you will need to sell 900 units to break even. Assuming you send out 10,000 direct-mail packages, you'd require a response rate of .9 percent. If you choose not to include a premium (a recipe booklet), you save $2,000, which reduces the break-even point to 800 units—a substantial change.

Of course, you also have to allow for returns—possibly 10 percent of total sales. That raises the break-even point by about 90 units, to roughly 990.

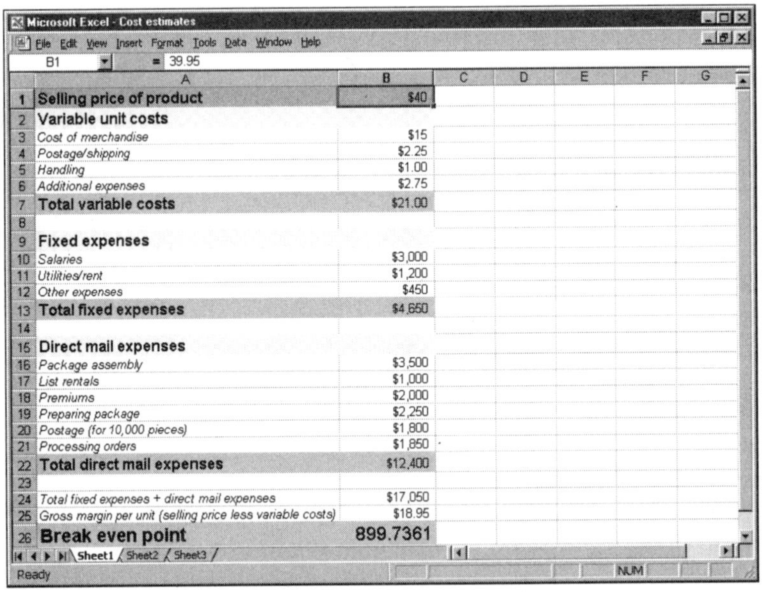

	A	B
1	**Selling price of product**	$40
2	**Variable unit costs**	
3	*Cost of merchandise*	$15
4	*Postage/shipping*	$2.25
5	*Handling*	$1.00
6	*Additional expenses*	$2.75
7	**Total variable costs**	$21.00
8		
9	**Fixed expenses**	
10	*Salaries*	$3,000
11	*Utilities/rent*	$1,200
12	*Other expenses*	$450
13	**Total fixed expenses**	$4,650
14		
15	**Direct mail expenses**	
16	*Package assembly*	$3,500
17	*List rentals*	$1,000
18	*Premiums*	$2,000
19	*Preparing package*	$2,250
20	*Postage (for 10,000 pieces)*	$1,800
21	*Processing orders*	$1,850
22	**Total direct mail expenses**	$12,400
23		
24	*Total fixed expenses + direct mail expenses*	$17,050
25	*Gross margin per unit (selling price less variable costs)*	$18.95
26	**Break even point**	899.7361

Figure 6-1

An Excel worksheet calculating the break-even point for a hypothetical direct-mail campaign.

Working with the Postal Service

If sticking a stamp on a letter is a DC-3, sending a mailing of 5000 pieces or more is the space shuttle. You enter a whole new world of complexity and regulation when working to meet postal service regulations. Particularly since the U.S. Postal Service changed its regulations regarding bulk mail in the mid-1990s, direct-mail marketers have found themselves increasingly responsible for ensuring that their mail is properly sorted, addressed, and bar-coded.

You have a surprising ally in this—the postal service itself. In most metropolitan areas you'll find a Postal Business Center (PBC). A good PBC (a lot depends on the local manager) can be an invaluable ally for a novice direct mailer. Its personnel can help with package design, list clean-up, even direct-mail techniques. "We can either give you direct advice or act as a middleman to get you in touch with people who can help you," says Tim Culver, a customer-service representative with the Seattle PBC. "We'll sit down with you and figure out what your best course of action might be."

In addition, the postal service offers an array of useful publications (all direct mailers should have a copy of *The Postal Service Guide to Direct Mail*) and an above-average Web site at *www.usps.com*. You can find the location of the nearest PBC at *www.usps.gov/busctr/welcome.htm*.

Of course, there's a reason why the U.S. Postal Service wants to be nice to you. They want you to do more of their work—or "work share," as Culver says. You can't blame them; the postal service handles 200 billion pieces of mail per year, about 800 pieces for every adult and child living in the United States. By ensuring that your addresses are "clean" and deliverable and that your mail is presorted, you can qualify for discounts of up to 46 percent off standard mailing rates.

Sorting Mail for Fun and Profit

In the mid-1990s the postal service and representatives of large mailers embarked on a new approach to classifying mail. Today there are three main categories: first-class, standard, and periodical. Standard mail, in turn, is broken down into subcategories, such as standard "A" (most direct mail) and standard "B" (parcel post). The postal service considers bulk mailings to be at least 500 pieces of first-class mail or 200 pieces of standard.

For either first-class or standard mail, you can earn discounts by taking several steps. For what is called "nonautomated" standard mail, discounts off the basic rate can be earned by presorting according to zip code or sending a larger number of pieces to a single zip code. For deeper discounts, mail can be "automated." This entails placing a bar code on each piece, which the postal service equipment scans and sorts by zip code. To automate your mail, you'll first need to check your lists against software certified as compliant with CASS (Coding Accuracy Support System), a postal service procedure for improving addresses and verifying zip codes by checking them against a national database. Your local Postal Business Center can supply you with a list of vendors that provide such software. If you find that more than 5 percent of your mailings are returned as undeliverable—as "nixies"—you also may need to process your mail in accordance with National Change of Address (NCOA) processing, a national database that the postal service maintains. Letter shops and service bureaus pay for access to the NCOA database and then sublet that information to mailers.

The last step will be to either box or bag your mailing, label the containers, and deliver them to a postal center or regional bulk mail center with documentation that supports the level of presorting you've accomplished. Your local PBC can supply you with the paperwork necessary to ensure you're in compliance with their regulations. A letter shop or full-service printer such as K/P Printing could also handle the details for you.

Paying to Mail Your Piece

You have several options for handling per-piece postage for a mailing. What you choose depends on your budget and your goals for the mailing.

> **Stamps.** The stamp is the most expensive form of postage, but it has the advantage of causing an envelope to look as if a human actually handled it—*it might not be a piece of advertising!* Some people in the direct-mail business take this idea very seriously; several really big direct mailers are said to have tested whether the response is better if the stamp is affixed straight or crooked. For standard mail you can purchase precanceled stamps to affix to your mailing. If you're using first-class postage, you can give your mailing a personal touch by using one of the postal service's wonderful stamps. Who could resist a mailing that sports a Daffy Duck or Tweety Bird stamp?
>
> **Metering.** Both first-class and standard mail can also be metered. Metered mail looks businesslike and professional. And recent

innovations, such as Pitney-Bowes' excellent Personal Post Office, make mail metering convenient for even a home-based business. The Personal Post Office is available with an electronic scale for weighing pieces up to 5 pounds, and it allows you to download postage by telephone. Once the postal rate is set, you simply feed in your envelopes. The Personal Post Office meters each piece and tells you when to insert the next one.

Indicia. Indicia are the preprinted bulk-mail permit marks you'll find on many mailings. Using indicia is the most convenient and inexpensive way to mail your piece; although the postage costs the same, you incur no labor costs for metering or affixing stamps. Your printer likely has a bulk permit and will allow you to use it for your mailing. Otherwise, you'll need to pay a one-time permit fee of $100.

Regardless of what method you use, postage is an up-front cost. If your printer or letter shop bureau is taking care of mailing your piece, you'll be asked to pay for postage before your mailing goes out.

Bulk Mail Facts

To qualify for discount rates, standard mail must meet the following requirements:

- All pieces must be in the same category (all letters or all flats).
- Each mailing must contain at least 200 pieces or 50 pounds of mail.
- The correct zip code must be on each piece.
- All pieces must qualify as third-class mail—no checks, bills, or other material that must be sent via first-class mail.
- A permit or license to pay postage via precanceled stamps, permit imprints, or meters must be obtained.
- An annual bulk-mail fee must be paid.
- Mail must be sent from the post office where a mailing permit or license was obtained and the bulk fee paid.
- Mail must be presorted.

Business Reply Mail

If you want people to respond to your mailing by mail, save them the trouble of fishing for a stamp. Many tests have shown that your mailing will draw a far higher response rate if you pay for return postage. To do this, you'll need to get a permit and a permit number. You'll also need to design your response card so that it meets postal regulations. There's a reason why all the response cards you see are postcard-sized—anything larger requires that you pay first-class postage rates.

If you expect relatively few responses—say, 500 to 600 pieces—you'll likely want to get a regular business reply permit. The permit costs $100; you'll also pay first-class postage plus 30 cents for each piece that is returned. You pay for the returned mail when it arrives. For more enthusiastic responses, you'll want to acquire an advance deposit account. In that case you'll pay in advance for the mailings you expect to return, and the postal service will then deduct the actual cost from that account. An advance deposit account requires a heftier up-front fee of $400, but you'll then be charged only 8 cents per piece in addition to first-class postage costs.

Mail Glossary

- **Address Change Service (ACS).** An automated process that provides change-of-address service to participating mailers who maintain computerized mailing lists.

- **Aspect ratio.** The dimension of a mail piece expressed as a ratio of length. For instance, a postcard that measures 5 or more inches long by 3 or more inches high has an aspect ratio of 1.57. An aspect ratio of 1.3 to 2.5 is required for automation capability and the accompanying discounts.

- **Bulk mail center (BMC).** A postal service center specially equipped to handle automated mail.

- **Coding Accuracy Support System (CASS).** A service offered to mailers, service bureaus, and software vendors that improves the accuracy of matching listed addresses to delivery point codes, zip+4 codes, 5-digit zip codes, and carrier route codes. Direct Mail Manager in Microsoft Office 2000 is CASS-compliant.

- **Drop shipment.** The use of a nonpostal service to deliver packaged bulk mail to a postal facility located closer to the final destination of the mail. Drop-shipping will qualify a bulk mailer for discounts.

- **Indicia.** Imprints on pieces of mail indicating that postage has been paid.

- **Permit.** Authorization to mail postage without a stamp, using indicia or other imprints.

Using Microsoft Direct Mail Manager

Office 2000 features a new tool for the small business—Direct Mail Manager, developed with Microsoft partner Envelope Manager Software. For mailings of up to 3500 pieces, Direct Mail Manager can import addresses from house lists in Microsoft Access, Microsoft Excel, or Microsoft Outlook; collect ODBC data from sources such as Microsoft SQL Server; or dial up third-party rental lists. In any case, it automatically creates a mailing list, checks it against an online CASS-certified database, checks for and removes duplicates, and prepares the mailing for delivery to the U.S. Postal Service.

To start using Microsoft Direct Mail Manager, simply click the Windows Start button, point to Microsoft Small Business Tools, and click Microsoft Direct Mail Manager. When Direct Mail Manager starts, you'll see that it operates like a Windows wizard, walking you through the steps you need to take to complete your mailing.

Your first step will be to select a data source. If you have set up a house list using Access, for instance, click the Browse button to locate the list from the files on your computer. Double-click the list to import it to Direct Mail Manager. You'll often find that the Direct Mail Manager readily recognizes your data format; for example, if you've created a house list using Access that lists customers by last name, first name, street address, state, and zip code, Direct Mail Manager will easily arrange that data into its proper mailing sequence. If there's any question, Direct Mail Manager uses a graphical map to double-check how you want data assigned (see Figure 6-2). Click Next when you're sure your data fields are correct. You can click Back at any time to change your decisions.

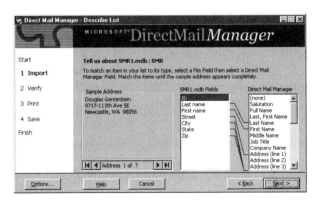

Figure 6-2

Direct Mail Manager's dialog boxes walk you step by step through the process of preparing a mailing for delivery.

Direct Mail Manager then allows you to select all or a portion of your lists, so if you want to restrict a mailing to a certain zip code or state, you can easily make that decision. Then Direct Mail Manager allows you to verify how the addresses will appear when mailed. Your screen will look similar to Figure 6-3. You can review as many addresses as you wish; if this is the first time you've used a particular list, you probably should take a look at quite a few.

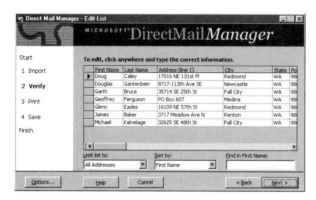

Figure 6-3

This dialog box lets you double-check how address material will appear on your envelopes or mailing labels.

After that, Direct Mail Manager will connect via the Web to a CASS-certified database called Dial-a-Zip that verifies addresses and postal codes and ensures that your list conforms to U.S. Postal Service standards. It then displays the entire list in a table with rows corresponding to each customer name and fields corresponding to the different address components. You can make corrections directly in the table, if need be. Direct Mail Manager also will hunt out duplicate names and give you the chance to remove them.

Finally, Direct Mail Manager presents your distribution options, such as standard mail, first-class mail, or mailing through a letter shop. You can change the design of the address printing, and then route the list through your printer to print on envelopes or address labels. To do that, Direct Mail Manager will open Microsoft Word 2000 and begin the print job. Then you can save your list, head for the post office, and wait for those returns to start coming in!

Chapter 7

Test for Success

Not long ago, California direct-mail consultant Jackie Walts Bailey was helping client Pacific Bell Information Services improve its direct-mail campaign to sell a voice-mail service. The company was facing declining response rates to its mailings and lower conversion rates when people who responded ultimately didn't purchase the service.

In most cases like this, says Bailey, a company will respond by sweetening the offer. "They'll offer a month of free service, then free installation, then both or more," she says. "We recommended they try offering a no-cost offer—a 30-day no-risk guarantee. They didn't think that would work, but we tried it and tested it."

To the mild surprise of the PacBell managers, the no-risk guarantee got the best response of several offers they tried. They never would have learned that had they not tried the no-risk guarantee, tested it, and compared its performance with that of other offers.

Testing is an essential part of direct mail. One of the great strengths of direct mail is that its results—unlike those of advertising that tries to build "good will" toward a product or has other hard-to-quantify objectives—can be precisely measured. You can establish with certainty that your mailing drew a 2 or 6 or 1 percent response rate and figure out how much it earned for you based on your printing, mailing, and fulfillment costs. Of course, that also means you'll know without any doubt when something flops. As Seattle direct-mail expert Bob Hacker says, "The good news about direct mail is that you can measure it. Unfortunately, that's also the bad news."

Because you can measure direct mail, you can test it very precisely. You can test to see if one list works better than another, what sort of letter copy is most effective, whether a pitch works best to generate leads or orders, or—as in the case of PacBell—what sort of offer is most appealing. Most veteran direct-mail practitioners test just as they brush their teeth—religiously and almost without thinking about it. "For those clients who let us, we never stop testing," says Hacker. Many small-business direct mailers, however, are content to play it by ear. They don't take the time to find out *why* a mailing does or does not work, or how seemingly small changes in it could boost response rates and profits. Instead they play it by ear—a mailing seems to work, or it doesn't. They may also devise packages based on their own biases. And that is a mistake, says Hacker. "The market will tell you what works," he says. You hear its words through testing.

What Does It Mean to Test?

Simply put, testing involves measuring a mailing's results and then trying to determine how you can make it more effective. You can begin testing with your very first mailing (and you should—try to make every mailing a learning experience). To do so, you might measure the results you get from different lists, test whether a 10 percent discount or a money-back guarantee is your best offer, or test whether a price of $24.95 or $29.95 results in greater profits. If you've already done a mailing and kept good records of how it performed, you can try to "beat" it by changing the offer or package to, say, 25 percent of the recipients of your next mailing.

How to Start

To be worth anything, a test must be precise. You need to know, for instance, that Package A drew a 4.6 percent response, while Package B drew 3.8 percent. You need good records of *how* those responses came in, whether by phone or by mail. And you need to be certain that your data is clean, that factors you aren't aware of or ignored didn't influence the data. Good testing takes effort, but poor testing isn't worth the bother.

Moreover, good records will allow you to learn as the years go by. You'll learn from past mistakes and successes, and you will become your own best teacher as to what will and won't work.

For starters, keep careful records of each mailing, says Jackie Walts Bailey. Save several copies of the mailing in a box or file folder, preserved exactly as they were sent out. Include in that file copies of receipts for all expenses associated with that mailing, including the costs of lists, printing, postage, and any incidental fees such as mail sorting. If you used the mailing to sell an item, keep records of what it cost for you to manufacture the item or purchase it from a wholesaler. And record what it cost to warehouse and ship the merchandise. In other words, your baseline information should enable you to determine precisely how much you profited from any one mailing and how much each sale and/or response cost you. If your mailing is a great success, make it easy to repeat by keeping the names of the printer and designer, or even a copy of the camera-ready art.

Next you want to start keeping track of those people who respond. Remember our discussion of a house list in Chapter 3; this is another chance to refine that list as well as to begin gathering important testing data. You want to know the following:

- The names of those who responded, along with their addresses, phone numbers, and other pertinent information

- What respondents bought

- How much they spent in aggregate and for each order

- How they responded—by phone, mail, or the Internet

- How they paid—by check, money order, or credit card

- On what date they ordered from you

- What mailing or offer they responded to (more on this later)

- If this is a business-to-business mailing intended to create leads, information about the contacts, their names and titles, any individuals above or below them in the corporate hierarchy, and what you need to do next to convert them to paying customers

A big part of this process, of course, is setting up a *method* for record keeping. Here Microsoft Office 2000 can be an invaluable ally. Using Microsoft Access, for instance, you can set up a database that contains all the pertinent customer information. Then you can easily extract entries based on particular criteria. If you want to target customers who have not ordered for a year, those who spend more than $1,000 a year, or those in a particular zip code or even carrier route, it's easy to do with Access. The key is to be certain that you maintain that database. If your business is very small, chances are your mailings also are fairly small. Set aside 15 minutes a day to input new responses. If you manage a growing small business, train a member of your staff to perform the data-entry task.

What to Test?

You can paraphrase most veteran direct-mail experts' advice about testing to a single phrase: Don't get cute. Sure, it's possible to measure responses based on whether you use red ink or blue ink on the envelope, whether you use a stamp or meter or indicia, and whether the stamp is on straight. "*Readers' Digest* can afford to test 25 different things with a mailing," says Ohio-based direct-mail consultant Dean Rieck. "But a small business usually can't afford that. So you want to test the things that really matter." For the most part there are two key factors to test: lists and offers.

First, Test Lists

"Getting the message in front of the right people is the number-one thing to do," says Rieck. Test as many lists as you can. True, you're paying for those lists and those charges can add up. But if you rent five 5000-name lists through a broker, try sending 500 mailings to every tenth name on each list (to ensure a random selection). Assuming that one or two of those lists look promising, based on the responses they elicited, you can then roll out a larger mailing to those complete lists. It may mean that you throw away several lists, but that's much cheaper than sending an expensive direct-mail package to people who don't want the darn thing.

Once you find a list that makes money for you, ask yourself a famous question from the movie *Butch Cassidy and the Sundance Kid*. Fans of that movie will recall that its heroes were being chased by a posse of faceless and nameless lawmen. After each dogged effort to lose their pursuers failed, either Butch (Paul Newman) or Sundance (Robert Redford) would ask plaintively, "Who *are* those guys?"

You, of course, don't want to lose your pursuers. You want to find more. But the question is a valid one. Who *are* the people on this winning list? What is their age? Income? What business are they in? Where do they live? Are they homeowners or renters? Male or female? Once you divine these answers, ask yourself what other lists include people with the same characteristics. Often the answer is not what you'd expect, says Rieck. He recalls a nonprofit client seeking child sponsorships who discovered that magazine ads run in teen magazines got a response as good as the response for those that appeared in such obvious publications as religious magazines. "Young women responded well to these appeals," he says. "But that's not something you'd think of sitting around a board room."

Let's say you're a software company that is coming out with a new flight simulator. You might assume that readers of gamer magazines would be obvious targets. Based on response cards from initial list sales, you might be surprised to find that many of your buyers are over 50, male, and either pilots or interested in things with wings. Where to find *these* guys? Rent lists from *Air & Space Smithsonian*; that's their classic demographic. And, indeed, the pages of that magazine are stuffed with ads for flight simulators.

Next, Test the Offer

Once you find a list or lists of people who seem interested in your product, you can fine-tune your offer. You can add a 30-day return policy, a premium, or a multipayment option. You can offer 2-for-1 or buy-3-get-1. For business-to-business mailings, you might test a written description of a product versus a demo disk of the same product. "There are all sorts of permutations on the offer," says Rieck. Look at what other direct mailers have done with their offers and see what looks appealing.

Of course, price is part of the offer, too. If your product is an item that already exists in the market, you'll probably want to base the price on what the competition is doing. That typically means trying to offer similar features and benefits for less money. But you never know—people often are willing to pay a premium for an item that they perceive to be superior *if they get a convincing sales pitch and the product warrants the price*. Your mission is to be persuasive in your communication with the customer and quality-driven with your product.

If you are offering a new item or service, who knows what the right price should be? Only by testing several prices can you determine the "sweet spot" that brings in the most sales with the highest possible price. One direct marketer recalls testing the same product at $249 and $149 and getting more responses with the $249 price! Why? Because the buyers seemed to have more confidence when they paid more for something. Had the marketer simply rolled out a mailing with the $149 price, thousands of $100 bills would have been sacrificed.

Test "the Creative"

Lists and offer/price strategies are the most important elements for a small business to test. But once you've built up a track record and believe you have the right lists and the right offer, it's perfectly OK—if you can afford it—to test your "creative," the letter or brochure copy and the package design. "Sometimes testing creative can save your backside," says Carol Worthington, a California direct-mail designer and consultant. She has seen creative tests beat a control by a ratio of 6 to 1. "Good direct mail is like a watch where every single gear needs to be working right," she says. "You need a good list, a good offer, and good creative."

How do you test something that seems subjective? Worthington recalls working on a direct-mail campaign for the Home Health Handbook, a card series written to help parents handle common family health problems. The offer was based on receiving a free notebook plus several card packs. To test the creative end of things, Worthington's group took three approaches:

1. In the first version of the mailing, the envelope posed four questions about family health and sex-related issues. It used questions designed to make recipients wonder just how well they could handle a family health crisis, such as "How can you save your partner's life with your own two hands?"

2. In the second version, a highly fear-driven approach, the copy posed a hypothetical emergency and asked readers if they knew how to respond.

3. The third version used a rational/cognitive approach based on the quality of the product and the strength of the offer.

The client and even the ad team were convinced that the second approach, trying to scare the pants off a parent, would be the easy winner. It wasn't. The first approach drew a much better response. "It had some elements of the cognitive approach, but it was more compelling and involving," says Worthington. "And it covered more topics than the 'emergency' approach, which dealt with a single medical situation. Had we chosen the 'emotional' approach—everyone's favorite—and gone with it, we would have had a 2 percent lower response rate and never known a better way."

There are many other ways to test the creative end of your direct-mail package, says Worthington. Among them are these approaches:

Using a plain white envelope instead of a colored envelope or an envelope with copy. At some point, many direct mailers say, you should always try a blank white envelope. The mystery it creates—even when the recipient is 90 percent positive it contains an ad—is one of the most effective tools devised for getting the envelope opened.

Enclosing a trial CD of a software product. Instead of using the mailing to drive people to your Web site, where they can download the software trial, include a trial CD to make it easier for recipients to actually install, use, and evaluate your product.

Trying different elements inside the envelope. You might, for instance, test your mailing with and without a "lift" letter. A lift letter is a copy piece that is slightly smaller than the standard letter, written to those people who think they have decided against purchasing the product. "We understand your skepticism," a typical lift letter will say. "But before you put down this letter I want to take the time to assure you that our product is as good as we say it is." You can test by removing almost any element from a package—omit the brochure, for instance, and use only a letter and a response form. Don't take out the letter, however; an unaccompanied brochure in an envelope is a "guaranteed bomb," says Worthington.

Testing premiums. For a mailing sent to owners of sport utility vehicles, you might test a piece that offers a handbook on off-road driving against one that offers a drain plug used to release dirty oil.

Testing how much information people want. You can test a long letter that says a lot versus a short letter that simply highlights key features.

Tracking Tips

Tracking your results is extremely important. Here are a some tips to help simplify the process:

- Encourage your customers to reveal information. If your direct-mail campaign is aimed at building traffic in your store, for instance, print a brief message on the piece that asks reader "to bring coupon in for an additional 10 percent off." The coupon will help you track responses.

- Have salespeople casually ask customers how they heard about your company.

- Code order forms with a letter code, number, or symbol that identifies the package that drew the response.

- Keep it simple. For most small businesses, tests of two or three packages at a time will yield more meaningful information than trying to test four or five different things.

The Right Way to Test

Direct mail should not be influenced by emotion, your gut feeling, or what "looks" right. It should be influenced only by what your tests show. To get information that will really improve your direct mail, though, you must be rigorous in your test procedure. In general, you should follow these four key guidelines:

> **Control the variables.** Make sure you know what you are testing. If you want to test both the letter copy and the offer, test them separately. If you want to test lists, use the same offer and package for all the lists that you test. Pitfalls await the unwary—you may think you're testing an offer, but find that because of a quirk in your list selection you are really testing geography. And beware what Bob Hacker calls the "invisible mutation syndrome," where many small edits and changes result in an entirely new package. Some common test strategies that control variables include the following:
>
> > Testing the same package against two different lists
> >
> > Testing different packages within the same list
> >
> > Testing a package with different elements, offers, or premiums within the same list
> >
> > Testing a list's "hot list"—the most recent responders—against a sample of the rest of the list
>
> **Test key details.** Again, few small businesses can afford to test a 10-percent cyan screen on an envelope versus a 30-percent cyan screen. Test what matters: list, offer, and key creative elements.
>
> **Keep track.** Devise a simple and reliable way to track your results, such as marking response cards with the codes "DM-1," DM-2," and so on, with each different package getting its own code. And make sure you keep up with the responses as they come in. Small mailings in particular are difficult to measure well because of the small sample size; the more responses you track, the more meaningful the result.
>
> **Establish a control package.** You're about to run a test, but against what? You need a *control package* against which you can test new lists, new offers, new copy strategies. If you have never worked

with direct mail before, your control becomes the first package you can send out. From then on, your objective is to beat that control package with a new combination of lists, offers, or creative. In each mailing, the control will make up a significant portion of what you send out so you can compare its results against the results of the new package. And don't let emotion rule. You may like the new package much more, but until it beats the control, your gut feeling is meaningless. You are, as Bob Hacker says, "a marketing survey of one." And one is a not a statistically valid sample size.

Bob Hacker's "Power*Test*"

It would be wonderful if all small businesses could afford to do mailings of 50,000, testing multiple lists and multiple offers before rolling out, but they can't. In fact, if you're planning your first direct-mail campaign you may not even be able to afford to collect a statistically significant number of responses.

That doesn't mean you can't test. Bob Hacker has devised what he calls the Power*Test*. Its goal: to gather as much information as possible in as short a time as possible and create a control that you can roll out with some confidence or use as a benchmark to run further tests. The objective of a Power*Test* is to develop a control package as quickly as possible. With a properly run Power*Test*, you can usually find a winner—or several winners—in the first test, and almost always find one by the second test. Here's how it works.

Phase 1

The objective of Phase 1 is to determine the key success drivers:

- Discover which lists will work for the offer
- Determine which offers will work for the lists

To show you how the process works, let's look at a 25,000-piece Power*Test*. The key assumptions are these:

- There are two package formats.
- There are three offer splits.
- There are five lists being tested, each with 5000 names.

Based on these test assumptions, the test matrix would look like Table 6-1.

Table 6-1 Testing Direct-Mail Offers

| Test | Offer A | | Offer B | | Offer C | |
Matrix	Package X	Package Y	Package X	Package Y	Package X	Package Y
List 1	833	833	833	833	833	833
List 2	833	833	833	833	833	833
List 3	833	833	833	833	833	833
List 4	833	833	833	833	833	833
List 5	833	833	833	833	833	833

In this matrix there are 30 test cells, each with 833 names. The results from each cell are not statistically valid. To generate a statistically valid sample, this test would have to include about 150,000 pieces, a number three times larger than the entire mailing universe of 50,000.

The results are highly indicative, however. You will see strong trends and "hot zones" that tell you what to do next. Let's assume that the test generated the response rates in Table 6-2.

Table 6-2 Response Rates for Direct-Mail Offers

| Response Rate Matrix | Offer A | | Offer B | | Offer C | |
	Package X	Package Y	Package X	Package Y	Package X	Package Y
List 1	4.5%	3.8%	6.4%	4.6%	2.5%	1.9%
List 2	3.4%	2.8%	4.4%	3.9%	2.7%	1.8%
List 3	1.6%	1.3%	3.2%	2.8%	1.6%	1.4%
List 4	1.7%	1.6%	2.2%	1.6%	4.5%	3.2%
List 5	1.4%	1.5%	1.5%	2.1%	1.4%	1.4%

In this example, assuming you need a response rate of 3 percent or more to hit the economic target, it is safe to conclude the following:

- Package X is a clear winner over Package Y.
- Offer B worked best with Lists 1, 2, and 3.
- Offer C worked best with List 4.
- Nothing worked with List 5.

Phase 2

In Phase 2 we check-test the results of Phase 1, this time using cells that are large enough to generate a statistically valid sample size. Assuming we could find a new list (List 6) that matches the profile of List 4 and can replace List 5, the check-test matrix might look like Table 6-3.

Table 6-3 Testing New Direct-Mail Offers

Check Test Cell Counts	Offer A		Offer B		Offer C	
	Package X	Package Y	Package X	Package Y	Package X	Package Y
List 1			5000			
List 2			5000			
List 3			5000			
List 4					5000	
List 5						
New List 6					5000	

With a response rate of 3 percent or higher and test cell counts of 5000, the results from the check-test would be statistically valid. You can find a winning control package with only 50,000 names, instead of the 150,000 that would have been required for a test of this magnitude using traditional test methods.

Testing Smaller Samples

Still, even a 50,000-name test is apt to be beyond the scope of many small businesses. If that's the case, talk to your list broker or manager about renting a sample list that contains 5000 names selected from a much larger list. This list typically is called an *Nth* list, says list broker Carol Kollman. That's a fancy name for a process of selecting every fifth or tenth or twentieth name to ensure a random sample. The resulting sample allows you to test several lists for a reasonable fee. If one list seems more promising than another, you can roll out your mailing by renting larger portions of the list or by renting several small sections from lists that are similar in demographic content.

How big a sample do you need to get meaningful test results? It's hard to say. "It depends on the product line, the price point, what the mailing is all about," says Kollman. "My usual suggestion is to send out at least three to four thousand." Of course, you can test this yourself. Over five years, you might send out several test packages to 2000 recipients. After rolling out to a larger audience, you might find that those 2000-package response samples accurately predicted the rollout results with a margin of error of, say, plus or minus 25 percent. If you're diligent in your data gathering and have a track record that gives you some confidence in the direct-mail game, you might find that you can devise your own standard for what constitutes a "statistically significant" test.

Common Testing Mistakes

Testing, says Dean Rieck, means sometimes having to live with results you didn't expect or didn't want. "Some delicate souls don't really want to 'test,'" he says. "They want to confirm." He lists several common mistakes people make when testing:

Testing haphazardly or running sloppy tests. "Testing is a mathematical process," Rieck says. "You have to test all the time. You have to test carefully. Otherwise, the numbers just won't mean anything. If you don't have the skill or patience for number crunching and analysis, get someone else involved."

Assuming that your tests are error-free. "Even if you run what you believe are careful, well-conceived tests, never assume that there is no room for error," says Rieck. "You should actively seek out mistakes on every level. Whether your test comes out good or bad, think through the whole process to track down errors. For a mailing you might ask: Were the mailing list numbers accurate? Were the addresses good? Did all our pieces get mailed? Was the bar code on the reply form correct? Are phone operators and mail handlers carefully tracking every response? Have I made mathematical blunders anywhere? Where else could a mistake be made?"

Drawing the wrong conclusions. "Too often, people look at test data and jump to a conclusion—'That self-mailer bombed. Self-mailers don't work!' or 'We tested a Christmas appeal, and we lost money. Christmas is a bad time to mail.' This is usually the result of a poorly designed test. Ideally, you should test with the express purpose of measuring one variable. And you must test against a proven control."

Making decisions based on insignificant results. "Every test must be statistically valid," says Rieck. That means you must reach enough of your audience to ensure that you have accurately sampled that audience and that you get enough responses to accurately calculate your results. "When you fall below certain minimums," Rieck says, "your results are worthless."

Filing away results instead of using them. "Why test if you just calculate a response rate and throw your report into a filing cabinet?" Rieck asks. "Those numbers are expensive to get, so use them. Analyze every test quantitatively and qualitatively. Show the number and write your thoughts and conclusions. Then share the test data with everyone involved. After every test you should know something useful like 'this two-page letter works just as well as this four-page letter' or 'this offer increased inquiries by 35 percent.'"

Failing to keep a running record of conclusions. "Over time," says Rieck, "as you see the results of test after test, you will start to see patterns emerge from the numbers. Organize and list this information as a guideline for future testing. Building on your hard-won knowledge will dramatically increase your success rate."

With every mailing, find some way to make it an educational experience for you and your business. You can become your own best expert on what direct-mail approach works best for you.

Part 2

Customer Management That Builds Sales and Loyalty

Chapter 8

The New World of Customer Service

Downtown Westport, Connecticut, used to be lined with small, locally owned shops. No more. Now national chains such as the Gap and Ann Taylor have elbowed them aside. With one notable exception: the clothing store Mitchell's of Westport. Founded by current owner Jack Mitchell's parents, Mitchell's of Westport has survived—and thrived—in large part due to its superb customer service.

"My parents started with the idea that they'd offer wonderful service and know all about each customer," says Jack Mitchell, who joined the store in 1969. "But as we grew, I noticed we were increasingly reliant on a woman named Olive. She did all the accounts receivable and kept all the information on customers on pieces of paper." Mitchell decided there had to be a better way, so beginning in the 1970s he automated Mitchell's of Westport, going through a series of machines until the company purchased a midsized IBM computer and hooked it to terminals on the sales floor.

Mitchell's goal: Use customer data stored on the computer to improve customer service and retain Mitchell's of Westport's retailing edge. As customers make purchases, Mitchell's 25 salespeople try to glean all the information they can. Buying habits, clothing and shoe sizes, color preferences, dates for birthdays, anniversaries, and other key occasions—all are stored on the company computer. "Rather than trying to remember every customer's name, we try to capture data on customers and use it in a way that benefits them," says Mitchell.

Mitchell's salespeople use the information constantly. When a familiar customer walks in, a salesperson will quickly call up that customer's records on a terminal screen, reviewing the customer's buying habits, size information, and price preferences. Or the salesperson will call to tell a customer that a preferred suit style is in. "And then," says Ray Cerittelli, who has worked at Mitchell's for 14 years, "I'll go to the computer and look up exactly what he bought in the past, his shirt size and shoe size, and when he comes in I'll have the suit fully accessorized. He can have a look at a complete wardrobe without having to spend a lot of time finding things."

Mitchell's is an upscale store, stocking high-end suits from Hickey-Freeman, Zegna, and other makers. The store makes certain its best customers are well cared for. Again, using the computer, Jack Mitchell tracks customers who spend more than $5,000 a year in his store, ensuring that they receive an extra measure of service. That's helped boost the number of big spenders who visit his store from 286 in 1992 to more than 1000 now.

But those are obvious sales. Mitchell's also uses its customer-service system to almost literally create business. Recently, for instance, the company created a list of all customers who had not made a purchase in more than two years and those who did not spend more than $900 for a suit. It then sent out 3000 direct-mail pieces to customers on that list, inviting them into the store with a special promotion. The result: 438 of those customers came in to make a purchase, generating $313,000 in sales. Mitchell's also can target mailings based on customers' preferences in clothing manufacturers, notifying them when a favored maker's trunk show is due into the store.

Such uses of technology will likely drive all successful retailing in the future, says Ben Barnes, IBM's general manager for global business intelligence. "We're using technology to get back to the idea of what a mom-and-pop store can do," he says. "But today we're dealing with maybe 50,000 or 70,000 customers, and you can't keep track of them without a computer."

Customer Service—A New Era

The example of Mitchell's illustrates a key point: Providing great customer service can mean the difference between survival and defeat for a small business. The small-business landscape is littered with the debris of startups that never made it past their second year; 20 percent don't make that cut, says the Small Business Administration. Many are obvious targets—sleepy, small-town businesses that were crushed when a Wal-Mart rose on some farmer's field a mile outside town, or businesses that fell prey to fickle consumer tastes. Others were seemingly well-run, successful businesses in major urban areas that couldn't cope with the new generation of "category killers"—big hardware or electronic superstores that offered both the lowest possible prices and a selection that smaller, local businesses couldn't match.

Mitchell's tale also says something else: The small business today has more tools than ever before with which to offer great customer service. That clothing store is a perfect example of what has been called the advent of "high-tech high-touch"—the use of technology to offer better customer service than even the most attentive human, working on his or her own, ever could. Mitchell's, Lands' End, L.L. Bean, and Amazon.com are just a few of the companies that have used today's technological power to provide a level of customer service not possible a decade ago. By deploying effective customer-service tools—and Office 2000 has a powerful one in its new Customer Manager—the small business can survive, and thrive, even in today's fiercely competitive market.

The Changing Competitive Environment

Ten years ago, business consultant Peter Glen wrote an angry book about customer service called *It's Not My Department!* In it, he described what he saw as an underlying problem in American industry: "Our products don't work," he railed. "We make cars that fall apart and sweaters whose sleeves come unraveled before we get them home from the store."

Fair enough, at the time. Anybody here remember the 1978 Chevrolet Vega? And do you recall the last time you saw one on the road? That was a car whose front end fell off during an early track test. The model never quite recovered; it was a clunky, wheezing, poorly designed rattletrap. It was no wonder that Japanese automakers made such hay in the United States during the 1980s.

Today, though, Glen would have a harder time making his case. Differences in automobile quality still exist, but they are much smaller than they once were. The global market forced U.S. businesses, grown lazy and fat after four decades of the most amazing seller's market the planet has ever seen, to relearn what it means to manufacture a quality product. Not long ago, for instance, *Consumer Reports* mentioned that it now typically finds two or three "sample defects" in cars it tests: misaligned parts, mysterious engine squeaks, a rear-view mirror that falls off. Even a once-maligned American automaker often will produce a car with zero or one sample defect. A decade ago, a score of *10* sample defects was considered exemplary.

This is a point that all businesses should note, says Dr. Kathy Leck, executive director of the corporate education division of The Lake Forest Graduate School of Management. She notes that successful companies must succeed in three arenas. One is product development—the ability to conceive of and create new products that the market will want. For decades U.S. automakers didn't really worry about that facet of business; they made the cars they wanted to make and let marketing do the rest. No more, as hot-selling models such as the Ford Expedition and Chrysler Concorde attest. Second, a successful company must have strong operating systems. Again, U.S. auto companies (well, GM still is having some problems) learned that lesson well, with even the Japanese now looking to companies such as Ford to learn how to keep costs down.

Third, says Leck, is the component that will make or break companies in the years to come: the ability to foster customer intimacy. Why will that one be so important? Because, says Leck, most businesses today have mastered the first two. "There really isn't that much difference today between consumer products," she says. "The ability to compete on quality or price has diminished, so customer management and customer loyalty become increasingly dominant." Take a look around you at the range of products and services available: big-screen televisions, personal computers, automobiles, graphics firms, restaurants, law offices. Chances are, a quality or price survey of any 10 vendors in the same industry would not reveal huge differences. Why are one or two firms or products in each group doing well? Most likely, because they offer excellent customer service.

Even that bar has become higher, says Leck. Blame it on Nordstrom, which years ago carved out a distinctive and hugely profitable niche for itself by selling more or less the same garments as everyone else and charging more for them. Nordstrom got away with it because of the special cachet its excellent customer service provided, along with the appealing swagger that combination gave its employees. The story goes that one woman was trying to get shoes fitted, but because of a minor foot problem she needed one size for her right foot, a different size for her left. They wouldn't be able to sell her mis-sized shoes as one pair, would they? she asked. "Of course we can," came the quick reply. "After all, this is Nordstrom."

Today, however, the Nordstrom model has become widely copied, with the result that even good customer service sometimes isn't enough. "That's no longer an add-on or a luxury," says Leck of traditional measures of customer service. "People expect it. I remember when you'd go into a restaurant and all you'd worry about was whether the food was any good. But now restaurants offer valet parking and all sorts of issues that have to do with access and comfort."

The "New Entitlement"

That's a term coined by Jack Burke, president of Sound Marketing in California and a man who frequently speaks and writes about customer satisfaction. He adds one more element to the increasingly challenging world of customer satisfaction: the need for speed. Just as "good" customer service and "good" quality have come to be seen as entitlements, not luxuries, so too has fast response become a given in most customer transactions.

Burke sees this "new entitlement" manifested in several ways. No better metaphor exists than the acceptance of overnight delivery from Federal Express, United Parcel Service, and other carriers. The World Wide Web also has reinforced it, with speedy e-mail and fast access to research now everyday tools for most people.

Some businesses have picked up on this. It used to take nearly a month to get a home mortgage, Burke says. Now a smart financial institution can give an answer in 24 hours. But many businesses still operate on "their" time, not the much faster customer time. Jay Goltz built a wildly successful and profitable custom-framing company by emphasizing what was for that industry blazing speed: one-week turnaround on framing a customer's artwork. "Most frame shops still take three weeks," Goltz says. "It doesn't take that long; the poster just sits in a drawer most of the time."

In many cases, says Burke, automation has given us the ability to meet the customer's expectation of speed. The question is: Do we?

What You Get and How You Get It

You may have decided that your company will succeed by offering the best possible product. And perhaps by dint of hard work and superior engineering you do just that. The business world, though, is full of great products that didn't succeed. Sony's Betamax is an oft-cited example; Apple's computers were heading that way. The product, however, is only part of the equation that results in customer satisfaction; it's the "what you get" part, says Leck—the part the deals with physical quality. To complete the equation a successful company must also deliver on intangible quality, on "how you get" something. Think of it as the difference between the broiled tuna with lemon sauce on your plate and the dining room around you. The fish may be wonderful, but if the service is slow, the atmosphere lousy, the tablecloth dirty—well, something is going to seem fishy, and it won't be the tuna.

Good customer service builds on a quality product by complementing that product and making the experience of receiving it more enjoyable. One of the best examples around is Harley-Davidson. That company was nearly out of business in the early 1980s, but now it has roared out of the corporate morgue to become a $1.8-billion company. How? True, Harley changed and improved its products, but they still aren't necessarily any better than what is offered by Honda or Kawasaki or BMW. The company also changed the way it delivers its products. Today buying a Harley isn't just buying a motorcycle; it's buying a complete *experience*. Harley stores today sell not only $22,000 motorcycles but also everything from fringed Harley lingerie to watches and hats. And the stores steep the customer in Harley culture. "We have a religion out there," a company executive told the *Chicago Tribune* not long ago. "And we have the most loyal customers in the world." Loyal not only because of what they got, but also because of how they got it. Product + experience = satisfied customers.

The Costs of Poor Customer Service

Plenty of evidence proves that "customer service" is a term that many companies regard as sufficient simply when said aloud repeatedly. Perhaps it's due to higher expectations, but surveys taken by the American Society for Quality show that customer satisfaction dropped from the mid 1990s until late in 1998, when it finally began to nudge upward in several categories.

Roger Nunley, founder of the Customer Care Institute, an Atlanta-based consulting firm, agrees that customer service is something that is promised more than practiced. "Companies say they stress customer service," he says, "but they don't really understand the value of it. People expect a lot more than they did five years ago, and while companies are improving, a lot of them haven't kept pace."

Indeed, it isn't at all difficult to find examples where a company that claims to stress customer satisfaction apparently needs a remedial course in what those words mean. To wit:

- A long-time customer of the Bon Marche, a large department store in the Pacific Northwest and part of the Federated Department Store chain, wanted to write a check instead of using a store charge card with an interest rate of 19 percent. She was asked to fill out a form to ensure that her checks would be readily accepted. She did, but the next three times she visited the same store she ended up on the telephone with the credit department, where someone wanted to know just who she was and why she was writing a check. "I finally took it to their vice president," the customer says. "We'll see what happens." After several weeks, she had received no response from the company.

- A customer who had spent well over $1,000 at the local retail outlet of Performance Bicycles, a national mail-order company, during the previous year brought a bicycle in for some quick adjustments. While a shop employee was working on the bike, the store manager wandered over. "Don't spend much time on that," he told the employee as the customer watched. "We have other customers in the store." The manager then wandered back over to the register to resume his conversation— with another employee.

- U.S. Bank sent credit customers a note that they'd better make their payments on time or they would face much higher interest rates. With the *next* bill, customers who hadn't met the "new" conditions during the previous year saw their interest rates jump from 11.9 percent to almost 20, with the promise that after a year their situation "would be reviewed." Not surprisingly, banks rate among the lowest in customer satisfaction, according to the American Society for Quality.

- Another customer stood patiently in line after Christmas to return a set of wine glasses to an Eddie Bauer store. As she did, her eyes fell on the prominently displayed motto, "Customer Satisfaction Guaranteed." Apparently, Eddie Bauer's definition of that term is fairly loose. The customer was not allowed to return the glasses— a gift—for a refund, only for store credit, despite the fact that the glasses were in a box bearing the Eddie Bauer name and each had an Eddie Bauer sticker. Fortunately, the gift-giver was the customer's sister, so without causing too much offense the customer was able to secure the receipt. Back at Eddie Bauer, she now was told she could not get a credit on her credit card because it was under a (duh) different name. No cash, either. Finally, the customer surrendered and took the credit, which she then carried around for weeks before finally finding something that appealed to her.

In all cases, the store that treated the customer as a nuisance rather than a valued client risked losing the customer, perhaps for life. Big mistake. Think of good customer service as a puffball, bad customer service as a rock. Which one can you throw farther? And how does the recipient feel when it arrives? Nunley recalls surveys taken during his days at Coca-Cola. "We found that a satisfied customer would tell maybe 4 or 5 people about their good experience," he says. "But a dissatisfied customer would tell 9 or 10 people." Think about it: If an angry customer goes out the door, that could well mean that 10 people will not come *through* the door during the next month or two. In fact, Nunley says, studies typically show that it costs twice as much to find a new customer—when you consider costs of advertising, sales calls, mailing pieces, and other customer-netting tools—as it does to keep an existing one. And in an age when price and quality tend to be uniform, keeping customers satisfied may be the key to keeping them.

How to Bug the Customer

It's really amazing to consider the indignities some people will suffer before they decide enough is enough. Most people who find themselves in the role of customer can identify a list of common mistakes companies make; some examples follow:

Phone mail hell. Using phone technology to cut costs is common, and most customers are not surprised if their call is routed to an electronic switching system. But how many option lists do you ask customers to listen to? And how long does it take for the typical customer to get to the correct party? If a customer has to hack through more than four layers of choices, you're not winning any friends. And, even though it costs money, let them "zero out" to reach a live operator by dialing "0."

Long wait lines. Today's rhetorical question: Why does a business exist? To make money. Who gives the business that money? The customer. Then why do so many businesses make the customer work to perform that essential function? Quality Food Centers, a regional grocery chain based in Washington State, uncannily sends additional checkers to the front of the store when lines start to build. But many grocery stores and other stores that feed all customers through a front-of-store check stand don't get it, requiring customers to wait for what seems like an interminable time to part with their hard-earned dollars. This makes no sense.

Oh, you called? When customers try to contact a company, they typically want some sort of response quickly. The problem is "top of mind" to them at that moment. Yet many companies see the customer on the company's own terms. They put the complaint or question low on the priority list, which results in a gap between when the customer wants an answer (now) and when the company intends to supply it (later, maybe *much* later). The worst culprits today are Web sites that invite customers to send in a question or comment via e-mail, and then fail to assign an employ to deal with the message in a reasonable time frame.

Helpless employees. Most people really don't like making a fuss. Yet if an employee is not empowered to deal with a problem and calls for a higher-up, then that is exactly what happens.

The Gains from Good Customer Service

Conversely, good customer service reaps enormous rewards for the company that gives it. True, as Nunley notes, it's bad news that, like the rock, tends to travel farthest—and hurt the most. Good customer service does indeed reap huge rewards. Says Leck: "You don't want customers to just use your service, you want them to *advocate* for it."

Most successful small businesses will attest to the power of that advocacy. Ron Musgrave runs Revive, Inc., a family-owned carpet-cleaning business based in Bellevue, Washington. Musgrave says that on average 17 percent of each year's business comes from customer referrals. Over the years, he says, that has translated to more than half his customer base. And it's about 20 jobs per month that come through the door at no cost to him, compared to the *single* job per month that is drawn by an ad in the yellow pages that costs nearly $400 *per month!*

Companies that spend less to attract customers can spend more to make their business better. They can improve products, train employees, increase pay or benefits. The result is a workforce that believes more strongly in the product and is more committed to the company. That, in turn, improves customer service by ensuring that customers see a familiar face when they come in the door, which, in turn, ensures more visits by those customers. And, that familiar face is likely to be upbeat and enthusiastic about the job.

Creating the Customer-Driven Company

How to avoid those pitfalls? By building your company around the customer, not around the boss or the employees. This is difficult, particularly for a small business. The entrepreneur can become so caught up in the hectic, day-to-day demands of running a young and growing business that he or she easily forgets that there's a world outside the office. But there is, and although it may not be a world (for you, at least) of golf dates and leisurely lunches and three-week vacations, it *is* a world where customers live. You need to ensure that your business inhabits that world.

What Does a Customer-Driven Company Look Like?

Mitchell's is a customer-driven company. Another is Douglass, Inc., a 38-year-old Indianapolis-based supplier of heavy equipment to construction and equipment-rental companies throughout the Midwest. Armed with laptop computers and customer-contact software from GoldMine, Douglass's salespeople dial in daily to the main office, downloading sales reports and outbound faxes to the company's central local area network and fax network while uploading e-mail messages, product data, or new information about a customer. "A single call brings them up to speed," says Douglass. Also, before making a sales call, Douglass employees can call up customer data to check year-to-date sales information and review a customer's history with Douglass.

Back at the main office, support staff is kept appraised of changes in a customer's contact information, credit history, equipment needs, and other information. "Now, when a customer calls in, 75 percent of the time the administrative staff can take care of what they need," says Douglass.

The result has been a boon for customers. "They're about the best" among the reps with which he deals, says George Marischen, vice president and general manager for Portman, Inc., an equipment dealer in Cincinnati. "They give us instant feedback on an order. With most reps I have to call the factory and wait a day or two to get an answer. But I can call the Douglass main office and they have the information I need." Adds Vito Addotta, president of Rock River Bobcat and Implement, Inc., in Rockford, Illinois: "When I have a question for our salesman, his office pages him and within an hour or two he calls me. With other reps, it can take days."

Mitchell's and Douglass, Inc., are very different companies with very different customers. But they share many characteristics of a customer-driven company. Kathy Leck, of The Lake Forest Graduate School of Management, says the customer-driven company will have the following characteristics:

A vision of customer service. Good customer service must be a bedrock foundation of the company, not an add-on. That vision requires that the company's executives develop a clear, understandable mission statement that underscores their customer-service goals.

A human connection. "Whether you achieve that on the Web or through face-to-face," says Leck, "the organization has to be customer friendly and customer oriented." That can mean different things to different companies. For Mitchell's, it means remembering a customer's spouse's birthday and knowing his or her shirt size. For Douglass, Inc., it means sending out reps who are prepared and responsive and don't waste the time of people who are wrapped up in running their own businesses.

A view that the customer is the top priority. Both Mitchell's and Douglass make the customer the number-one consideration, and they have standards in place to ensure that customers always receive the treatment they expect.

Uniform customer service. This trait is particularly important for larger companies, but it applies to smaller ones as well. What it means, says Leck, is that regardless of what division or store or employee a customer is dealing with, the experience will be similar to other experiences with the same company. "There's a common way that the customer gets listened to, or how they're communicated with," she says. "Geography and other factors will cause some variation, but the flavor will be the same."

The ability to learn. A customer-driven company constantly seeks to change itself *based on customer feedback*. This goes beyond tracking things such as sales growth. That is a good indicator, but a drop in sales may show up months after a customer begins to be dissatisfied with a company's performance. To learn from its customers, a company must constantly think of ways to measure how satisfied they are and then take steps to put that knowledge to work. And it must be prepared to learn from its competitors or from other businesses. So what if you didn't come up with the idea? Every success in the world has been built atop someone else's success—or near miss. Anyone who doubts that should read Richard Rhode's *The Making of the Atomic Bomb.* The book clearly and painstakingly shows how a series of brilliant scientists, like a precise basketball team moving across time rather than space, passed ideas from one to another over 50 years to come up with the gadget that ended World War II and changed the world.

Knowledge of the customer's world. Whether it's driving into a fast-food restaurant because they're hungry or buying a $5,000 copier to smooth document handling, people take action to solve a problem. Companies that give great customer service *understand the world of the customer.* That means they understand why that customer has turned to them. In the business-to-business world, for instance, customers have clients of their own, and they are turning to you to help them better serve their own customer base. If you let that customer down, you potentially disrupt his or her own business.

An ear attuned to the customer. A customer-oriented company listens to what the customer has to say, says Leck. It regards complaints not as criticism, but as a chance to improve the company. The listening company makes it easy for the customer to have that say. This can be tricky in most of the United States— outside of New York, certainly. Restaurant people will tell you that Americans tend not to complain about a dish; they just eat it in silent disgust despite the fact the maître d' has the power— and often the willingness—to fix the problem. Moral: You have to encourage customers to speak up and speak their mind.

Hiring and training policies that encourage customer service. Customer service starts with the interview. So hire right, and hire with customer service in mind, and you'll do far better in the long run.

Empowered employees. All companies say they "empower" their employees, but do they really? Empowered employees are given both the tools and the confidence they need.

In the following chapters, you'll learn how to use this model for a customer-driven company to create a business where the customer is the focus for everything you do.

Microsoft Customer Management Tools

Microsoft Office 2000 is a valuable ally as you work to become a customer-centered company. Most of its tools you already know about and already use—Microsoft Word, Microsoft Excel, Microsoft Outlook, and more. New in Office 2000 Small Business, Professional, and Premium, however, is a tool specifically designed to help give great customer service: Microsoft Small Business Customer Manager.

What Customer Manager Can Do for You

Microsoft created Customer Manager with the idea that most small business owners had plenty of valuable information about their customers, but rarely were able to use it to its full potential. To turn that information pile into an information mine, Customer Manager creates a database of accounting and contact information that can be used to analyze business trends and opportunities.

When you first start Customer Manager, you will be asked to pick an existing database with which to work or to start a new one. Databases are saved in Microsoft Access format, and they can contain accounting data from third-party business programs such as Intuit's Quicken as well as contact information from Outlook. Now you can use this information to create datasheets or "hot reports" that identify your best customers, determine order status for an individual account, track e-mail messages you have sent to a client, coordinate correspondence to customers with e-mail or Microsoft Word, set up automated reminders for sales calls or other customer contact, and more. With Customer Manager you'll be able to analyze your customer relationships in detail, so you can spot noteworthy trends and leverage them to your advantage.

As you learn more about great customer service in the chapters that follow, you'll see how Customer Manager can help you deal with specific customer management issues.

Chapter 9

You and Your Customer

Customer service often does not get the attention it deserves. Why? Because at many companies it is an intangible. Good customer service—and the customer loyalty and profits it creates—isn't easily calculated in a spreadsheet as sales and advertising costs and new-product development expenses are.

Every company, though, should make the effort to value and nurture excellent customer service. By understanding who its customer is, a company can then write an effective vision statement that outlines its commitment to customer service.

Customers Come in Different Flavors

The term *customer service* may conjure up a democratic ideal about everyone's being treated the same, all equal in the eyes of your business. Well, that's baloney. Take a look at back issues of *National Geographic* and find a picture of a bunch of lions. Chances are, in the foreground of the picture you'll see a tasty little animal that you'd think the lions would cheerfully grab as a snack, but they don't. Why? Because it's not worth their effort. They're waiting for something *big* to come along.

You should not be quite that Darwinian, but the point is the same. If you frantically chase after every bit of business that wanders through the savannah grass (that is, your front door), you'll soon exhaust yourself and your employees. The fact is that for any business about 20 percent of the customers generate 80 percent of the revenue. Even though all customers deserve good service, it's silly to think that the 20 percent who do the most for you don't deserve something extra. They do.

Most businesses serve two kinds of customers. *Final* customers are those who buy and use your product or service. They come to your florist shop to buy a bouquet, hire you to write press releases, and count on you for on-time tax preparation. *Intermediate* customers are distributors or other business people who take your goods or services and resell them—perhaps to final customers. The line between them isn't perfectly clear, of course, but if you run any sort of retail business that's open to the public, you generally serve final customers. If you are more business-to-business oriented, many of your customers are likely to be intermediate customers.

So what? For starters, you need to know who it is you need to keep the happiest. In some cases intermediate customers are your only customers, or the very important ones. In other cases, even though you may see your intermediate customers more often, you might have final customers as well. If you sell a product to retail stores, those stores are intermediate customers. Your final customers actually *use* those products, and if the products prove unsatisfactory, you'll lose those customers. Although you want to pay attention to your intermediate customers, their concerns (credit terms, prompt delivery) may be different from those of your final—and ultimately more important—customer (good quality).

Larger businesses must deal with a third customer classification: *internal* customers. The public relations department of a corporation, for instance, works with outside news organizations or stock analysts. That department must also work with the executive suite and other divisions within the company. Here again, the question is: Who is the most important customer? It might be hard to persuade the CEO that he or she is not the final customer, but for the sake of the company it may be necessary to do so.

Define Yourself as Your Customer Defines You

Of course, your customers have their own view of you. Depending on the role you perform for your customers, they could see you as any of the following:

- **Vendor.** A vendor supplies a commodity product to customers. Chances are they view you primarily on a best-price basis, so your challenge is to devise a way to make them notice you for more than simply the penny a pound they save by buying from you. If your customers are homebuilders, and if you're selling two-by-fours, then those builders view you as a vendor.

- **Supplier.** Let's move a little up the food chain here. The customer values you as someone who offers a particular service or product that is not readily available elsewhere. You may supply homebuilders with custom roof trusses or with skilled finish work. You have to work to maintain that relationship, though; if you're not careful, the competition will figure out a way to turn you into a vendor.

- **Partner.** "Partner" confers top status. The builder now calls you before even breaking ground on the project, consults you about the proposed plans, and gets your input on the type of materials to use and where they can be most readily sourced. You even know the customers of the builder.

How do you see your business in relation to your customers? If you're a vendor or a supplier, is there a way to move up to the next level? There probably is, and probably the way to move up is through great customer service. Begin the vision-making process by seeing your company as your customers might. Why? Because nobody cares what you think about your company or its products. They care only about how you can help *them*. Have an out-of-body experience and picture yourself as a customer of your company. And start asking some questions of yourself:

- What does your company have that a customer would want?

- What is the benefit of your product or service to the customer?

- Why would a customer choose to deal with you instead of a competitor?

- How will customers know that you care about them and their business?

- What would drive a customer away from your business?

- What will the customer say about your company to other potential customers?

- How well are you serving the needs of the customer?

All Customers Are Not Equal

Remember, we talked about the fact that all customers are not created equal. And they aren't. Understanding that and acting on that principle allow you to serve your good customers better and improve your bottom line. Fair enough?

You'll find that customers generally fall into these categories:

- **Top customers.** These are people who view you as suppliers or partners. They are a minority in terms of numbers, but they account for the bulk of your revenue. If you know their names and can describe in detail their business or who they are, then they're probably your top customers.

- **Customers with potential.** Are there people or businesses that you would like to add to your list of top customers? Maybe a business that now deals with a competitor of yours might be willing to switch teams. Or, you have found that your top customers share certain characteristics, and you suspect that you can identify others who also share these traits.

- **Regular customers.** You see them fairly often—five or six times a year.

- **Occasional customers.** These people may order or buy from you once or twice a year, but they are not big money makers. In fact, they may even *cost* you money, when you factor in the administrative costs of carrying them in your customer records, sending mailings (if you do), and stocking items only they may purchase. Moreover, these customers may be among your biggest headaches. Remember that 20 percent of the customer base leads to 80 percent of your sales? Well, it's also true that 20 percent of your customers will cause 80 percent of the service calls, complaints, and other customer-service problems. It's surprising how often that group falls into this "occasional" category.

Within these groups, of course, you might find other ways to segment your customers. These might include the following factors:

- Geographic location
- Business type
- Size of business
- "End" versus "intermediate" customer

126

- Income level
- Age
- Product categories they buy

Use Microsoft Excel to create tables that can show you which customers are most valuable to you. Calculate, for instance, the total dollar volume each customer brings you, how much per contact or visit that customer spends, your estimated expenses for maintaining contact with him or her, and other factors.

The Butterfly Customer

Susan O'Dell, a Toronto-based consultant who has worked internationally with clients on customer-service issues, coined the term "butterfly customer." She notes that customers once were bound by geography and transportation limits to a small area. Neighborliness and brand loyalty were the order of the day.

No more. Today technology, changing demographics, and a suburban culture that has lost its sense of "neighborhood" have produced a customer who will bounce from store to store, Web to street corner, mall to catalog. O'Dell and consulting partner Joan Pajunen have identified eight traits of butterfly customers.

1. **They accept your invitation to be loyal.** Butterfly customers will happily sign up for your customer-retention program—and those of three or four of your competitors. They may indeed be repeat customers, but are they *loyal* customers?

2. **They move across market segments.** Stores used to be known by their customers. No more. The same customer will stop at Nordstrom and K-Mart during the same shopping trip; he or she will spend $500 for a Sony DVD player and visit Target to get the best buy on a DVD disc. Similarly, customers who might have seemed "down-market" once now will surface to snap up a $35,000 Dodge Durango.

3. **They are intelligent and well educated.** More information is available now than ever before—and lots of people are taking advantage of it. Customers may now know as much about your business as you, or they are willing to learn. With the Web, customers can know what a shopper is apt to pay for the same item six states away—and they want the same price locally.

4.   **They are skeptical.** Too much information, ironically, also makes consumers wary of what they're promised.

5.   **They would rather switch than fight.** Surveys show recent improvements in how people see customer service. But O'Dell says that improvement is also due to the fact that people don't believe they're heard. The danger here? You may be about to lose a customer—and not even know it.

6.   **They listen to their friends.** And now, with the Internet, they have zillions of 'em. Via chat rooms and newsgroups, Web-oriented consumers (a fast-growing number) now have easy access to total strangers who nonetheless can strongly influence their buying decisions. This development could ensure that bad news about your company travels far faster than it did just a few years ago.

7.   **They're not embarrassed to be butterflies.** People used to see the local shopkeeper in church or at a restaurant; no more. People have no feelings of obligation toward those with whom they have no social contact. And they wear this freedom on their sleeve. Butterfly customers will set merchandise down and walk out of a store if made to wait too long, or they won't hesitate to ask for a discount for cash.

8.   **They know their value.** Butterfly customers, says O'Dell, are all too aware of your desire to make them regular customers. And they're ready to remind of you of that if you don't recognize it.

OK, Now What?

Once you get a grip on how your customer base is divided, you can then figure out how to capitalize on that knowledge. The short answer is this: *All* customers deserve high-quality, attentive service. In fairness to your business—and your best customers—make the experience of doing business with you particularly easy and memorable for a select group of customers, your customer "A" team. It's not uncommon, for instance, for companies to assign customers to three tiers—Gold, Silver, and Bronze. Use whatever labels you wish. Right at the top with your best customers, says Roger Nunley, you should include those customers you would *like* to have as "A" customers. Into the middle tier go your regular-but-not-best customers. And into the lower tier go those who purchase from you only occasionally.

How do you give those top-ranked customers special treatment? There are plenty of ways, says Nunley. Among them are these:

- Assign a special account representative who calls on the best customers. Customers further down the hierarchy get sales calls once a month, instead of once a week (or on whatever schedule you think appropriate). The rest of the time they use a toll-free number to order from a Web site.

- Offer a discount to top customers only, or send occasional mailings that give notice of special sales or offer special financing.

- Let top customers use a special phone number that sees a lower call volume than your general sales number. Got someone on the staff with a great telephone personality? Put him or her on that line.

- Tailor billing procedures to mirror a top customer's own accounting procedures, such as when that customer likes to pay his or her bills.

Remember, you don't want to give certain customers *inferior* service. Always keep in mind that an infrequent customer may simply be testing your business before committing to becoming a major customer. By the same token, a small customer today could be a growing business. He or she might have the potential to do two, three, or even four times the current volume of business with you.

Creating a Customer-Service Vision

Now that you have a grip on your customer—a gentle grip, of course—write a statement that expresses your goals for securing and keeping that customer's business. We're talking about a "vision" of customer service. True, "vision" is much overused these days. Any customer-service person you talk to, though, will tell you right up front: If you're going to excel at customer service, you need to say so in writing. "It's the first thing you should do," says Roger Nunley, founder of the Atlanta-based Customer Care Institute. "And if the CEO is not on board and not supporting that customer-service vision, then it's not going to go anywhere."

But why bother? you might ask. I'm in business, and I know I need customers, so doesn't that by definition mean I care about the customer? Hardly. A vision statement is your road map for the customer care component of your company. It's the core value for your company, the statement that will guide it when questions come up. And it will be the glue that binds your company together, particularly if customer service is a function performed across several divisions or offices. Otherwise, you may discover that each semi-independent or independent unit is making up its own rules as time passes. Given the natural evolution of things, you might find that your Duluth office and your Pasadena office have developed completely different, and perhaps even contradictory, views of customer service.

Moreover, a vision statement will serve to inspire a company's employees—and leaders—in a way other things cannot. People want meaning in their lives, and they want meaning in their work. By creating a vision that your employees can rally around, you create an intangible motivator that reinforces and builds on the tangible motivators (salary, benefits) you provide. An excellent vision of customer service will challenge employees and staff, and it will give them a mission in their day-to-day work.

A vision statement that revolves around the customer also serves as an operational guide. Whenever a customer is angry, whenever your company makes a mistake, whenever you reach a fork in the road, the vision statement should be what guides you. Consider the oft-cited example of Audi when it was confronted by angry customers who claimed that their cars were suddenly leaping backward at high speeds. Audi engineers ran tests and determined that the cars were not at fault, that people were mistakenly pressing the gas pedal, not the brake, when they were backing up. Audi simply denied that there was a problem. And, technically, they were largely correct. The company, however, had forgotten the difference between *what customers got* and *how they got it*. The car was OK, but the delivery (blaming the customer, in effect) was wrong. And Audi paid dearly; between 1985 and 1987, sales fell from 74,000 cars a year to 26,000.

Involve Everyone

A vision statement for customer service needs the input and involvement of those will be most affected by it: your customer-service employees. "That helps achieve buy-in," says Roger Nunley. "If people have input they feel they were part of the process and will take ownership of the result." So get your entire customer-service team together—or key representatives from it—and have them work to develop their own versions of the statement. Ask them to list what they see as the main objectives of good customer service within your company and to develop a clear, simple statement that supports those objectives. Make it a brainstorming session—encourage lots

of rough drafts. This is a good chance to use the "mind map" discu
Chapter 4; ask people to write "customer service" in the middle of
of paper, circle it, and draw "branches" out from that circle. On each
write a component of customer service and perhaps methods for
ing it on subbranches.

Here are some other questions to pose when writing a vision statement:

- What is the core mission of the company? Why does it exist?

- What do people who deal with us now think about us? (Go ahead; call them up and ask.)

- What can we do that will distinguish us from competitors?

- How will we achieve our objectives?

The Three "C"s of a Great Vision Statement

What constitutes a good vision statement? Clarity, credibility, and cre-ativity. It should be as brief as possible and easy to comprehend. Avoid jargon and the current crop of buzzwords ("synergy," "pro-active," "best practices"). Make the vision statement credible with your employees; it should be believable (not "to give the best customer service the world has ever known!") and reflect their wording and values as expressed in focus groups held to discuss the statement. And although your statement won't win bonus points for style, a little verbal pizzazz won't hurt. It will make the statement more memorable.

Don't be shy about talking with other businesses about their own customer-care statements and looking to ones you like as models. Here, for instance, are some samples:

To build and deepen our relationship with our consumers, providing service that exceeds expectations and strengthens our heritage of trust.

Simple and to the point, this vision statement doesn't win many style points, but it has the ring of honesty and the flexibility to adapt to chang-ing conditions.

Our mission is to build and strengthen lasting relationships with each individual contact. We will provide timely, friendly, courteous, and expert service. Meeting or exceeding these expectations will help us become the "Amazement Company."

Maybe it's a little wordy, but this is another credible, clear statement. It emphasizes the importance of building customer service one customer at a time, and it seeks to add a memorable phrase in the objective of building the "Amazement Company."

Now, What to Do with It?

Writing a vision statement is like joining a health club. By creating that statement you've "joined" the club. But if you want to meet your goals (lose weight; improve customer service) you have to *use the equipment!* The vision statement must become a part of the company culture. That begins with the CEO, founder, or president—the top dog in the company. If the boss doesn't buy in, no one is going to buy in.

Then you must find a way to permeate the organization with the message. Some companies print their statement on small cards that can be handed out to and kept by employees. Certainly, posters with the message should be prominently displayed. And the statement should be repeated at every possible opportunity. Integrate it into performance evaluations and into job descriptions. You know you're getting through if these conditions exist:

- Employees can quote the statement, or at least the gist of it.
- The statement is used as a touchstone when problems are addressed.
- The statement is seen as a guide for employees' day-to-day activities.

Finally, remember that it's a vision statement, not a fossil. Make it an annual habit to reexamine the statement to make sure that it reflects your current goals and that is not going stale. Let the vision statement reflect and address your growing company's changing needs and customer base.

Setting Standards for Service

One you've determined how to segment your customers, it's time for the next step: What would your customers consider great service? It's important to have *measurable* standards; otherwise you'll have no way of knowing whether you're making progress. Those goals must be *achievable* as well. People who believe they are aiming at a receding—or worse, a moving—target are likely to get discouraged.

Standards will vary depending on your business. Many companies could follow the example of L.L. Bean, the Maine-based outdoor-goods catalog company. Bean makes customer service a near-religion, and it rigorously strives to set standards against which it measures its performance in customers' eyes. Bean's standards include the following:

- **Quality.** The company pays close to attention to what is being returned and why.

- **Availability.** Are advertised products in stock?

- **Service.** How many rings before a telephone is picked up? How many callers abandon the call before it is answered? Bean carefully tracks these measures. Its call pick-up average of only a few rings is an impressive standard in an era when many calls to high-tech companies result in holds of 20 minutes or more.

- **Packaging.** Most Bean products are sold via mail order, so the company works to ensure that its packaging is attractive and durable.

- **Employee knowledge.** Bean wants its telephone representatives to be as knowledgeable as possible about its products—how to care for them, how they fit, what they're best suited for. Employees are trained intensively in the Bean product line.

- **Retail service.** The company's loyal customers rightly regard Bean's retail store as something of a temple. When they come for a visit, they are given customer service that's hard to match.

But one L.L. Bean standard should be common to anyone who is interested in attracting and retaining customers. "We want every customer to feel they are being treated with the respect they deserve," says Elizabeth Spaulding, the company's vice president for customer satisfaction. "It's as if they are honoring us by making a purchase, and we are very appreciative of that."

Come to think of it, how many companies have a vice president whose sole task is to ensure customer satisfaction? Creation of that title alone would probably send many companies well down the road to improved customer service.

Barriers to Good Customer Service

In Chapter 8 I discussed some of the things that can really bug a customer. Usually, though, these problems are simply the product of a system that is poorly designed. They are, in other words, no more than the result your customer-service system was designed to produce. If you think about customer service in those terms—as a physical product—you can begin to see more clearly why a service *system* is important. If you want fewer defects in a product, whether it's a color television or good customer service, you need to know where to find the problems on the customer service "assembly line."

In the case of customer service, these "manufacturing glitches" can include the following:

- **Lack of a motivation system.** True, it's an employee's job to provide the best possible customer service. Why not reinforce the behavior of those who truly meet that obligation? Cash rewards, extra time off, recognition from higher-ups, prizes—any of those might be useful tools for rewarding great customer service. But make it meaningful; an "employee of the month" parking space is a hackneyed device (unless, of course, parking really *is* at a premium). Find out what employees really value in terms of recognition from you, and base your reward system on that. And make sure the behavior that earns that reward is clearly defined and understood by all—simply rotating the "prize" in the interest of fairness will breed cynicism.

- **Poor infrastructure.** If you want your employees to manufacture great customer service, what tools do they need? Do they have them? Whether it's a comfortable and easy-to-use telephone system, state-of-the-art PCs, or a well-organized stock room, make sure that the physical tools you provide to your employees send the signal that you care about customer service as much as you say you do.

- **Inadequate training.** Many times customer service "training" means showing a new hire the restroom and lunchroom, sitting him or her down next to a veteran employee, and leaving the new hire to absorb information by osmosis. Be sure you devote time to the technical and people-skill training that a customer-service employee needs to succeed.

- **Understaffing.** Remember from Chapter 8: Speed is the new entitlement. People don't like standing in line to make a purchase or have their questions answered. Usually, when they are asked to do so it is because too few employees are present to handle the current workload. And by putting more pressure on the employees on duty, you're apt to discourage them while making a customer's contact with that employee less than ideal.

- **Making it difficult to buy.** Sure, you want to sell something to your customers. But how easy is it for them to complete this transaction? Make sure your store or office is easy to find (have fax-ready directions); have enough staffing to speed transactions; make sure you have the credit arrangements customers need and expect. Always remember, it should be incredibly easy and painless for people to give you money. Don't make them work to do that.

- **Overpromising.** You want to be enthusiastic about your company, your people, and your products. But you want to be cautious about promising a customer more than the company can reasonably deliver. If you do overpromise, you're setting the stage for a disappointed—and perhaps angry—customer. Follow this axiom: Underpromise and overdeliver.

Superior customer service is not an accident. It's the result of good planning, of understanding your customer, and of establishing goals for excellent customer service. Take the time to develop a vision and commitment to customer service, and your customers will become your own best advertising system.

Chapter 10

Hiring and Training a Customer-Oriented Staff

After Jay Goltz graduated from Northern Illinois University in 1978 with a Bachelor of Arts degree in accounting, he opted for a slightly different career path. Instead of going to graduate school or getting a job with one of the big accounting firms, he drew on part-time work he'd been involved with for years and opened the Artists' Frame Service. He started in his dad's basement, then moved into the old factory district in Chicago.

Today, the Artists' Frame Service is the largest retail custom picture-framing company in the United States, with more than 120 employees and $9 million in sales. Goltz has written a book based on his experiences running his business, *The Street-Smart Entrepreneur: 133 Tough Lessons I Learned the Hard Way*, and he teaches what he calls "Boss School" seminars to would-be entrepreneurs.

Goltz, a fast-talking 43-year-old, built his business on a solid foundation of customer service. "I called it 'taking care of the customer,'" he says. "I thought that's what everybody was doing. But I learned that most companies *don't* take care of the customer. They say they do. But what I found is that they'll say,

'Sure we give great customer service. Well, except for that woman there, who's a pain in the neck. Or for that guy, because we lose money on him.' And that's how they slowly lose their customers."

Goltz finds all sorts of ways to ensure that customers get service they can't find anywhere else; you've come across his name elsewhere in this book. One of the things he believes is most critical to success is this: He hires carefully and takes pains to train his people thoroughly.

That principle is followed less often than you might think. During the late 1990s, for instance, employers everywhere said that there wasn't much they could do about ensuring that they hired the right people. In a booming economy, you often heard, it just wasn't possible to adhere to the hiring standards you maintain when times are tighter and the pool of candidates is larger. Goltz wouldn't hear of it, though; he simply worked harder to find the people he needed. And he takes a firm approach toward ensuring that his employees uphold his ideas about good and bad customer service.

Customer-service specialists across the board agree: It starts with hiring the right people. That means setting high standards for hiring. It means understanding what it takes for a person to succeed in customer service. It means hiring people with the right characteristics to succeed. And it means giving them the training and tools so that they can do their utmost to help you build your business.

Find the Great Customer-Service Employee

No matter how much training and motivation you apply after you make the hire, if you don't have the right kind of person to begin with, the training is apt to fail. And that can be enormously costly to your company.

Begin the hiring process by asking a simple question: What kind of person do I need for this job? If you are hiring for a job that emphasizes customer service, you should take a different approach and seek different qualities than if you're hiring for a sales job. People who excel at customer service have many skills. They also have a particular spark that sets them apart from other personality types, a spark that says they genuinely care about other people.

The Values Factor

Giving good customer service is not like fixing a fuel-injection system or writing software code, however. It's more than a set of skills; often it's also a *value*. Values aren't instilled in college classes, and chances are you can't do much to change them after you hire someone. Values, though, have a tremendous amount to do with how customers *receive* your product, and how they feel about the experience. Even if the meal in a restaurant is perfectly prepared or the newly purchased gift neatly wrapped, the manner in which the employee served the food or wrapped the gift influences how the customer feels about the product— regardless of its quality. Here are some elements of values:

- Honesty
- Friendliness
- Motivation
- Integrity
- Reliability
- Selflessness
- Diligence
- Care
- Conscientious attitude

Traits of Customer-Service Winners

Susan O'Dell is the owner of a winery near Toronto. She also is an international consultant on the retail and service industries and the author of *The Butterfly Customer: Capturing the Loyalty of Today's Elusive Consumer* (John Wiley & Sons, 1997). Although many characteristics can describe the effective customer-service employee, she says, two stand out: curiosity and a desire to be of service to others.

Curiosity

People who aren't curious, she says, live in the past. That means they'll evaluate their job and their customers based on *past* experience. People who are curious are constantly asking questions, seeking new information, learning about the world around them. "Although often more difficult to manage, curious people are worth their weight in gold," O'Dell says. "Turn them loose on customer research, and they will come back with information as valuable as that from an expensive study."

How do you find curious people? They'll often reveal themselves— people who are curious ask questions and are apt to make for lively interviews during the hiring process. To draw the candidate out, says

O'Dell, ask questions that begin with "Tell me about a time when…" Ask why certain procedures are followed in the candidate's current job or create a situation that requires curiosity to solve and ask how he or she would respond.

You win, I win

A second trait, says O'Dell, is a desire to be of service to others. "It's the joy of problem solving, meeting a need, or offering a kindness to a stranger," she says. During an interview, watch for a person's listening skills and the credit he or she gives to others, says O'Dell. See if the candidate takes pleasure in having overcome obstacles or solved problems in previous tasks. And see if he or she genuinely seems to like service work.

How to Find Great Customer-Service People

To find people with the right values for your business, you need to look past formal qualifications or the accomplishments listed on a resume. Human resource experts and executives who run smart, customer-oriented companies say one way to start is to establish consistent guidelines for hiring. These include the following:

- **Know what you want.** Your customer service vision statement should serve as your first guideline for your hiring practices. Next is the opening for which you're hiring. Describe to yourself in as much detail as possible what that job entails. Write a description as well of a perfect employee for it. Look around at your current employee base. Who are your best customer-service workers? What sets them apart from your other employees? Is it their ability to handle multiple tasks? To endure stress? To stay calm in the face of angry customers? Try to devise interview questions that will allow you to see if the applicants have these attributes. And use the key words and concepts from your prehiring needs assessment in any ads you run to help ensure that you attract the right candidates.

- **Look to the past.** People don't change that much. "It's axiomatic," says Monica Griffith, a human resources consultant. "What people have done in the past will indicate what they'll do in the future." So do what you can to find out how applicants have behaved in past jobs and past situations. It may mean that you call three previous employers for references.

They may be apt to open up, and by talking with several of an applicant's past employers you'll get a better idea of his or her behavior. Take a look at the applicant during the interview as well. Ask questions that will yield information about how the applicant reacts in particular situations. If you want your customer-service people to use their own good judgment to solve a problem, pose a hypothetical question about a situation the applicant might face. If his or her answer is a variation on "Well, first I'd consult the manual," that candidate may not be right for you.

- **Watch them work.** Companies such as BMW, Nucor (a steel firm), and Cessna put applicants into a simulated assembly line or even a simulated office to see how they do. This approach may not be practical for a small business. But if an applicant still has another job, deputize someone from your firm, someone the applicant won't recognize, to give him or her a look.

- **Hire people your people know.** Do your utmost to get referrals from those who already work for you. People tend to hang out with people who are like them; if you have friendly, competent people working for you, it's a safe bet they know other friendly, competent people. Train your existing staff to be your first-line recruiters, and reward them when their referrals result in a stable, productive hire.

Hiring Is a Verb

Jay Goltz believes that hiring the right people is a matter of looking for them. Hard. "It's too easy, in a tight market, to settle for less," he says. "But hiring is 75 percent of the customer-service game. You have to put out the right bait." He runs ads until they get results, and if they don't get results quickly, he changes the ads to see if different copy works. Once respondents come in, they get what might best be called the Goltz Treatment.

Size them up

Goltz takes a good look at new applicants. He wants to know their current job status—whether they're looking for a job because they don't have one or because they want to move up in the world. "Most people who are looking for a job because they're out of work aren't going to be a good fit," he says flatly.

Ask tough questions

Goltz doesn't believe in letting applicants off easy during the interview process. "If I ask them why they want to leave their current job, and they say, 'Well, I think it's time for me to move on,' I don't buy that," he says. "We go for the jugular. We ask them what they don't like about their current job, what they'd like to see that was different about it, whether they talked to their boss. I've never had someone who wouldn't open up once they're pressed a little."

Many employers interview candidates poorly. Most—especially new entrepreneurs—tend to focus too heavily on skills. Judging a good customer-service candidate is a tricky business because so many skills may come into play. You might ask "Have you ever had a grouchy customer?" That question is closed-ended, though—the answer might simply be "Yes." Then what? Instead, ask open-ended questions, such as "Tell me about the different customers you've dealt with" or "Tell me about your most difficult customer."

Interview Guidelines

Interviewing anyone is an art form, not an interrogation. If you make people comfortable and pose the questions carefully, you'll be amazed at how they will open up and be candid and forthcoming about themselves. Some tips:

Give the interview structure. An interview should have three parts. First, set aside some time to make the applicant comfortable and to make small talk. Give the candidate a chance to look you over even as you look him or her over. Next, get to the heart of the interview. What are the main things you need to know about the applicant? Finally, try some open-ended questions to get the candidate talking. Get the applicant to tell stories about memorable people or crucial decisions he or she has made. You'll learn lots about the candidate and his or her character.

Silence is golden. Don't feel that you have to keep the conversation constantly moving. After an applicant offers an answer to a question, fill the interval with silence. The candidate might feel a little nervous about the silence and will likely add more detail to the previous answer or perhaps say something he or she hadn't quite meant to say.

(continued)

Interview Guidelines *(continued)*

Get specifics. Framing a question as a response to a statement just made by an applicant is often helpful: "Could you give me an example of what you mean?" "Why do you say that?" "I'm sorry, I don't quite follow." Such questions delve deeply into what someone is thinking.

Be consistent. Have a basic question list that you ask all applicants. And don't let an applicant be passed from person to person, each with his or her own set of questions. Some companies go so far as to assign a single person to do the interviewing and tape the interview for others involved in the hiring process to review.

Get references

Nowadays a lot of employers discount references. They figure that in today's litigious climate an applicant's former employer will likely offer the careful answer rather than the meaningful answer. And that's true—up to a point. To Goltz, the lukewarm reference is as telling as the poor reference. He looks for managers who will really sing the praises of an applicant. "Every great employee I ever got had a great reference," he says. And he asks straightforward questions of an applicant's former employer. "I ask if the employee quit or was fired," he says. "The answer to that tells me a lot."

Training a Great Customer-Service Staff

CGI Consulting Group, based in the Philadelphia area, is an 80-person company that handles the service-sensitive job of managing company benefits programs. Its list of firms that have "outsourced" this task to CGI includes Benetton Sports Systems, Entercom, and Deloitte Consulting.

CGI places a premium on great customer service. "We have to be better than a company's own benefits department," says Harriet Hankin, president of CGI. Her measure: the "wow" factor. "When someone calls our 800 number with a question, we want our response to be so great that when that person puts down the phone they say, 'wow,'" says Hankin.

Getting that reaction, however, requires a superlative staff. CGI hires with immense care, with a highly formalized interview approach and a clear idea of the kind of person who will work well in the company's environment. Then new hires are put through a rigorous training process. "It starts with what we call 'Benefits 101,'" says Hankin. "We cover the basics of the benefit programs on which they'll be working." That's followed by intensive training on CGI's computer system and on telephone skills. Then the new hire works with a "mentor" employee under the direction of a supervisor. "They aren't on the phone for months unsupervised," says Hankin. Even veteran employees are given regular refresher courses, and daily meetings update employees on new developments in benefit programs and alert them to "hot cases" the company is handling.

Training of the intensity CGI offers is something few companies do, says consultant Susan O'Dell. Why? Because money spent on training often is difficult to track. Instead, many companies prefer to invest heavily in brick-and-mortar assets—new stores or new computer systems. Those sorts of investments are much easier for a company to measure, says O'Dell. A company can see where the money has gone and even budget for it in years to come based on maintenance or replacement cycles. Not so with the human element. O'Dell has even been told by clients about companies that avoid training an employee until he or she has stayed with the firm for three months. Otherwise, the company fears, an employee will take his or her new skills to a competitor.

Moreover, says O'Dell, some companies have come to believe training simply doesn't work. That's probably due to poor training or training that wasn't reinforced. It's also common for employers to forget just how much they themselves have learned in recent years. The new employee, though, doesn't have the benefit of that experience. "The employers sometimes have to rein in expectations," says O'Dell. "They have to tell themselves, 'The fact that I know this doesn't help the employee.'"

It's the task of an effective training program to convey that hard-earned knowledge to a new employee. A basic training program, says O'Dell, must accomplish the following:

- **Ensure that employees don't "practice" on customers.** In a sink-or-swim approach, it's typical for a company to simply toss a new hire into the tank. "The day before Christmas I'm willing to be waited on by anyone who will take my money," she says. "But the rest of the time I want to feel that the employee knows the job." Ideally, a company could consider investing in a mock business environment, putting employees through as many scenarios as possible. Add real-life pressure to the situation by timing an employee or throwing in unexpected complications.

- **Separate knowledge from skill.** It's a common mistake to forget the difference, says O'Dell. Someone may be able to describe how to make a serve in tennis, but *performing* that skill is a different thing. "It's the difference between talking about it and doing it," she says. An employer should think about what *skills* are needed to execute a job—whether it's to help a customer set up a PC or project a calming presence over the telephone—and train people to execute those skills. Often that means carefully breaking the skill into its separate components. If an employee needs to know how to calm an angry customer, for instance, what are the steps to that process? First, it's probably crucial for the employee to stay calm. Then an employee must coax the customer to say what he or she believes is required to fix the problem. The employee must then understand that information, temper unrealistic expectations, and resolve the problem. But how?

- **Follow up.** Many companies view training as an "inoculation," says O'Dell—the employee is given it once, and then pronounced "trained." But teaching someone a new skill requires reinforcement; by O'Dell's estimate, 80 percent of all training dollars are wasted due to lack of follow-up coaching and repetition.

Training for Maximum Performance

The steps listed previously are an employee's basic training—boot camp, if you will. A company that's truly devoted to getting the most out of its employees will go beyond the basics. O'Dell says that successful service-oriented companies combine three key elements: commitment to vision and culture, knowledge about the business, and knowledge about the customer.

Commitment to vision and culture

Your vision statement for customer service should be the first thing employees hear about, says O'Dell. You must convey your company's core values to a new hire immediately, within the first few hours of his or her tenure. Retrofitting those values after an employee has learned something else is difficult.

Knowledge about the business

People need to know the why behind the what. If you want employees to believe in your company and commit their efforts to it, says O'Dell, you need to tell them all you can about your business and their role in it. Once employees fully understand the implications of their work for the success of the business, they are much more likely to "buy in." Some entrepreneurs might feel a little queasy about opening the books for employees, but in many cases that is exactly what you should do.

Knowledge about the customer

Just as the business owner needs to see his or her business through the eyes of the customer, so should the employee. O'Dell tells of a gasoline retailer who dispatched new employees to a women's lingerie store. The lesson? The new hires soon learned what it was like to feel uncomfortable in a strange environment. As a result, they were able to take steps to make their own customers more comfortable. Other successful service companies regularly send their employees on shopping expeditions—with company money—so that the employees can experience being a customer in different stores. The information and tips they bring back quickly make up for the expense.

Training for the Job

Jay Goltz makes a further point about training: You should think about the particular challenges of a job and train specifically to meet those challenges. "Training usually isn't as hard as it appears," he says. "When you think about it, how many ways are there to screw up? Usually it's not 100; it's more like 10." Identify those potential problem areas and train to meet them. "If you say, 'OK, if this product is broken then here's what we do,' that's something an employee can stick in their tool belt," says Goltz.

Motivating Your Customer-Service Team

No one should suggest that your employees are worthy of comparison to a cocker spaniel, but here goes: Valerie Robbins, a dog obedience instructor in Seattle, tells of a client whose dog finally learned to "sit" when greeting a visitor instead of jumping up. "But these people make me so mad," the client told Robbins. "The dog is doing exactly the right thing, and they ignore her!"

The moral? It astonishes me how often the only behavior that gets any sort of reinforcement is the wrong behavior. It's the "problem" employee who gets the extra attention of an employer, and perhaps even a reward when the problems are solved. The best employees, meanwhile, often go about their business unseen and unrecognized simply because they're doing exactly what the employer expects.

The Motivational Triangle

A lesson in management can be learned from *Truman,* the wonderful biography of Harry Truman by historian David McCulloch. McCulloch recounts what took place when Truman was given command of an artillery company that was known as a bunch of misfits and malcontents. In the weeks and months that followed, Truman turned a gang that had vowed to ignore him into a military unit that was devoted to its new captain and soon showed itself to be the best artillery company in the regiment.

The key? Good leadership. Truman proved himself to be tough but fair. He brooked no lapses in discipline, but he was fair and human. He made sure his soldiers got good food and were spared the petty humiliations that army life can entail. It's a lesson any manager should learn: Peak employee performance begins at the top. By laying out the rules, setting a strong example, and expecting excellence you can motivate a winning customer-service team, just as Harry Truman motivated a superlative military unit.

Motivation, however, takes effort. A good manager understands that much of his or her job is to properly motivate employees and takes steps to ensure that motivation happens. Three key elements lead to effective employee motivation: rewards, recognition, and responsibility.

Booty

Everyone likes to feel that his or her best work will result in something more than the usual paycheck. Make it a habit to offer financial rewards for excellent customer service. The reward can be a cash bonus or desirable items such as gift certificates to a fine local restaurant or tickets to a concert. Be sure the employee who receives the reward knows what earned it for him or her, and try to give the reward shortly after the exceptional act of customer service you're acknowledging. Also, separate "team" from "individual" performances. For the team, reward the entire group by having pizza brought in for lunch or taking everyone to a baseball game. Groups can be tricky, though; you don't want to make the reward (such as an evening or weekend function) turn into an extra work-related obligation.

Also, be wary of contests that pit employees against one another or that reward the wrong behavior, such as processing customers *quickly* rather than *effectively*.

Way to go!

One of the most powerful yet least-used forms of motivation is simply a kind word. People crave recognition for their efforts yet so rarely receive it. And few things are more demoralizing than feeling proud of your work and getting no—or worse, perfunctory—acknowledgment. Be generous with praise.

- Post letters from grateful customers.

- Ask an exceptional employee to talk about his or her techniques to the team or to a different department.

- Send a handwritten note thanking an employee for good work. For greater impact, send it to his or her home.

- Include anecdotes about an employee's excellent work in a company newsletter or internal e-mail message.

- When you see good customer service being given, comment on it immediately.

Set them free

Perhaps the most powerful motivator of all is responsibility. It's amazing what resources and energy people will call upon when they know the game is theirs to win or lose. If you believe you've hired right and trained properly, don't hesitate to give people responsibility. How? Here are some tips:

- Authorize your employees to do what is necessary to satisfy a disgruntled customer.

- Show that you trust them. Give them new responsibilities, ask them to run a meeting, deputize them in your absence.

- Make them problem solvers. If there's an area in which you'd like to improve service quality or boost sales, give an employee or employee team the task of finding a solution.

The L.L. Bean Approach

L.L. Bean, a company renowned for its customer service, is careful to put all three elements into place. Bean is the nation's leading outdoor-related catalog company, with more than $1.2 billion in sales and a volume of more than 50,000 calls per day. Yet it works to ensure that each customer feels that he or she is doing business with a small company. A big part of achieving that effect is ensuring that its employees are uniformly well trained—and motivated.

Employees at L.L. Bean are regularly rewarded for giving excellent customer service. Individuals who have been singled out for excellence are allowed to choose from several bonuses: a choice of work schedule for the next three months; a parking spot near the door (worthwhile during the brutal winters in Freeport, Maine); a $100 L.L. Bean gift certificate; or a day off with pay. That system ensures that the reward is something meaningful for the employee, yet not so great that it sparks resentment among other employees. Teams are rewarded with ice cream parties or barbecues.

Recognition also is awarded liberally. As at most other companies that do a great deal of business over the telephone, L.L. Bean has managers who listen in on conversations between customers and company representatives. When a representative does a particularly good job, a manager immediately walks to his or her station and sticks a note that says "great job on that last call!" on that representative's terminal, even as he or she is speaking with another customer.

More formal recognition is given at regular ceremonies honoring employees who have given outstanding service to both external and internal customers. Candidates are nominated, and nominations are reviewed by managers and employees. The recipients are named at monthly meetings for which the company even pulls people away from phones. "We might take a hit in service for 10 minutes, but it's worth it," says Elizabeth Spaulding, L.L. Bean vice president for customer satisfaction. Spaulding herself makes every effort to attend each ceremony and name the winners herself.

Good performance also is recognized with added responsibility. Individuals or work teams, for instance, may be assigned to tasks such as "call-backs," in which they telephone customers whose ordered merchandise is out of stock to explain the situation and offer another product. Customers aren't happy to hear that there's a problem, of course, but most are thrilled to hear from the company, turning a customer service loss into a break-even proposition or even a win. Plus, the job gives the employees a chance to do something other than wait for their own phones to ring.

Is the Customer *Always* Right?

One of the worst enemies of outstanding customer service is cynicism. And one creator of cynicism is that old warhorse, "The customer is always right."

Basically, that's bunk. And it can lead to situations where your employees come to believe they're at the mercy of the irate, irrational, unreasonable customer.

Jay Goltz's approach is refreshingly human and humane: "Ninety-eight percent of the time, the customer *is* right," he says. "In those few cases where they're wrong, I tell my employees that it's just not worth arguing about. I tell them, 'Let's just *pretend* they're right.' Once you tell the employee that, it's kind of a relief to them. They can buy into it, as opposed to that line about the customer always being right."

Goltz also has come to the conclusion that a tiny minority of customers simply aren't worth having. "Is it OK to have a customer screaming, 'You stupid asses! I can't believe you did that!'?" asks Goltz. "It won't happen a lot, but I think it's important to have a discussion with employees about just how much they have to take."

In cases where a customer is deemed too painful to keep around, the temptation may be to say "Take a hike, pal," but it's best to be tactful. Try a brief conversation or a letter that says in effect "It appears we simply cannot provide a level of service you expect. We don't want to lose you as a customer, but perhaps you'd like to try Shop Y." Keep it low-key and nonhostile. Who knows—by handling a hostile customer with kid gloves you might ultimately win a more loyal customer than you could imagine.

Taking Tough Measures

While keeping impossible customers may require harsh action, so too does dealing with employees who simply aren't measuring up, says Goltz. "That's something you don't hear when you go to these business seminars and people talk about how nobody is an employee anymore, they're all 'associates' and they all 'empowered,'" he says. "But there has to be accountability. And that means someday you have to get tough and fire somebody."

That's something Goltz won't hesitate to do. He recommends that a business owner regularly take a hard look at his or her workforce. What that owner will see are four employee categories:

1. **Motivated and capable.** These are employees at the top of the game. You've hired well and trained well, and their internal values have done the rest.

2. **Motivated but not capable.** These are probably new hires who haven't yet learned all the ropes. You need to make sure these employees make it into the first category.

3. **Unmotivated and not capable.** It's time to act fast. These people are doing your company no good. Get rid of them.

4. **Capable but not motivated.** These people are management's biggest headache. These individuals may show flashes of Category 1 behavior, but they usually don't. Meanwhile, they can damage morale and injure your relations with customers. Figure out a way to move them into Category 1, or tell them to hit the road. "You have to say to yourself, 'If this person is not helping me make this a great company, then I have to let them go,'" says Goltz. "And don't let attrition do the job for you—if you wait until people quit to replace them, you're just breeding mediocrity."

You may not like it, but there are times when you simply have to act in your company's best interest. And that means asking someone to pursue career opportunities elsewhere.

Showing the Way

It is far better, though, to be in a position where you hire well and train well, and the result is a customer-service team that makes you and your company look wonderful. And remember: Great customer service starts at the top. Says Larry Kroin, founder of a highly customer-oriented company that distributes high-end kitchen and bathroom fixtures: "A store is a reflection of its manager. If I don't set the right example, nobody is going to follow."

Chapter 11

Keeping the Customer Satisfied

In some ways, the numbers are reassuring: Most measurements of customer satisfaction show that people today seem a little happier with the products and services they receive than they were a few years ago. Those numbers, though, tell only part of the story. People seem to be unhappy about plenty. Airlines, banks, cable companies—those are just a few of the industry groups that have been the target of consumer wrath in the late 1990s.

The target of complaints also has changed. The three industries mentioned in the previous paragraph are all service industries—not manufacturers. A decade ago, were you to survey people on their pet peeves, many of the answers might have been about shoddy quality: a tape player that jammed; a car that rusted in two years. The quality of most consumer goods has improved markedly as companies have upgraded manufacturing processes, and the electronics revolution has made many consumer goods simpler, more reliable, and more feature-rich. With many goods comparable in quality, consumers now are turning their attention to service issues. And, in many instances, they are finding that service is lacking.

There's another reason to believe that survey numbers regarding satisfaction may be misleading. Russ Merz, an analyst with the CFI Group, a

Michigan-based consulting firm that helps large companies measure customer satisfaction, notes that people often claim to be much happier with a product or service than they actually are. "People want to feel good about what they did," says Merz. "So when they look back at a decision, there's this need to say to themselves, 'I did it, so it must have been a good decision.'" There's even a term for that phenomenon—"attributive reasoning."

If customer satisfaction is, in fact, a house of cards, that spells considerable danger for many companies. Merz notes that 20 years ago about half the stock price of a public company was based on that company's physical assets—inventory, factories, real stuff. Now physical assets account for only about 20 percent of a company's capitalized value. The rest? Intangibles—including the goodwill with which the public and stockholders regard the company. "Service," he concludes, "has become increasingly important."

The Web and the Angry Customer

Figures concerning the impact of a dissatisfied customer don't take into account the technological nuclear weapon ticked-off people now possess: the Internet. Word of mouth now is global on the Web, and customers who don't believe a company has treated them properly do not hesitate to take advantage of it. One of the most prominent recent cases, of course, is the flap over the mathematical problems unearthed in an Intel Pentium chip released in 1997. Intel denied there was much of a problem, and the wired academic world made the company pay dearly for its mistake.

Since then the reach of the Internet has tripled, and the range of complaints people are taking to the Web has expanded as well. No one knows that better than Dunkin' Donuts. When customer David Felton grew angry because he couldn't find his favorite milk (1 percent) for his coffee, he didn't just get mad. He got a domain name (*www.dunkindonuts.org*) and started to collect complaints from other customers about everything from stale coffee to poor customer service to fresh paint spilled on a pastry.

To its credit, Dunkin' Donuts has rolled with the punches, tracking Felton's site closely, following up on (reasonable) complaints, and bombarding people who post on the site with coupons for free donuts or contrite e-mail. Moreover, Dunkin' Donuts is hardly alone. Nike, United Airlines, McDonald's, CompUSA, and the Gap are just a few of the companies that now are targets of "anti-company" Web sites.

(continued)

The Web and the Angry Customer (continued)

What to do if your company becomes the target of a Web-rage site? Start by listening to what those on the site are saying and ask the simple question: Do they have a point? Even though Web sites are easy to set up and not especially expensive ($100 to start, plus $50 a month to maintain), few people will go to the trouble unless they are *really* annoyed. Then treat the site founder as you would any angry customer. Talk about the problem, figure out how to resolve it, and move on. It may be an irritation, but such a site also can work in your favor. People have forgotten how to complain (or they have stopped trying) through regular channels—that they choose to do so on the Web is apt to yield some valuable intelligence you never would receive otherwise.

As a result, it's more important for a business to listen for —and even welcome—the sound of dissatisfaction. Better to hear a little grumbling now and again than to hear nothing at all. Of course, determining what you want to listen for can be difficult. Customers speak in many ways, with many voices. In this chapter we'll cover three basics: identifying customer types, learning to listen, and establishing basic survey techniques for hearing your customers.

Why You Want Complaints

How many times in the past week or month have you been disgusted by slow service, annoyed at a store employee's lack of attention, or put off by a lengthy hold on the telephone? Several times, at least. And how often did you complain? Probably not often; perhaps never. If so, you aren't alone. In fact, according to the Technical Assistance Research Programs (TARP), a research group that focuses on customer behavior, only about one out of every 27 people who are upset about poor customer service speaks up. Most believe it will be a waste of time and simply increase their irritation.

That is not to say that dissatisfied customers don't complain. As noted in Chapter 8, bad news travels fast when it's about your company. In a comfortable atmosphere, an unhappy customer will tell 8 to 10 people about his or her irritation. TARP has found that some soreheads will tell as many as 20 people.

At that point, the math gets pretty frightening. Many businesses—especially small businesses—rely heavily on word-of-mouth advertising. That can backfire in a hurry. If you hear a complaint about your company, you can safely assume that another 26 people have the same gripe. Each of them has told perhaps 10 people, perhaps more. Very quickly, that *single* complaint becomes evidence that *260 or more* people are either unhappy with your company or have heard about a problem with your firm. And once bad news gets out, it is very difficult to overcome.

You want complaints because they will help you improve your business. Complaints tell you something important—they give you a specific way to improve your business and your service. Moreover, you are apt to find that many complaining customers actually are loyal customers who want to continue doing business with you—provided you shape up. And, ironically, if you can make good on a customer complaint, you stand to benefit more than if you had provided good service in the first place. TARP has found that an angry customer, once satisfied, is apt to tell five people about his or her positive experience. Only three people will discuss a positive experience if the initial experience is good. Go figure.

Why Don't More People Complain?

You'd think it would ease the collective blood pressure of the nation if people would complain. But they don't. Why not? In most cases, people don't complain because they feel it's just not worth the trouble. They've learned over the years that when it comes to customer complaints, the burden of proof often rests on the complainants. They must provide documentation. They may be treated rudely or brusquely. They're apt to be blamed for the problem. They may be subjected to an interrogation the likes of which even Ken Starr couldn't conceive, given lengthy questionnaires to hand out, told to call someone else.

In short, they don't complain because most businesses make it difficult, if not downright unpleasant, to do so. Try this for a radical concept: Encourage complaints. Do away with red tape when a customer has a problem he or she wants resolved. Train your employees to listen to complaints graciously. Don't ignore customers who appear angry; approach them and talk to them. Let people talk, and chances are, they will. Your business will be the better for it.

What Customers Hate to Hear

Here are some examples of the corporate approaches that discourage people from complaining:

- "**It's our policy …**" Policies are shields you hide behind. Your customers need you to be flexible. They see themselves as individuals, and they want you to do the same.

- "**The computer was down …**" Computers don't make decisions; people do. Take responsibility, or offer to research a problem rather than blaming the error on "the computer."

- "**They …**" As in "They require that we …" This approach signals to customers that you refuse to take responsibility. If you do not have the authority to help a customer, locate a manager who can.

- "**I can't …**" Rephrase a statement such as "I can't handle this problem right now." Instead, say "I will handle this problem as soon as we finish."

DiSC: The Four Personality Types

Handling complaints requires a little psychology. It helps to understand what might seem obvious: People are different. Keep in mind, though, what that really means: Many people are different from *you*. Your style of complaining and responding to complaints may be very different from that of others. Handling complaints successfully involves understanding what kind of people you're dealing with and adapting to the style of the person doing the complaining. Says Kathy Leck, executive director of the corporate education division of The Lake Forest Graduate School of Management, "Knowing your own communication style and how to adapt it to the kind of work you're doing is an important part of customer service. Knowing your own style and that of others also lets you communicate well when you're under stress, which is often the case when there's a customer problem."

People aren't cut from a pattern, of course, but most people embody one of the four styles summarized by the acronym DiSC: Dominance, Influence, Supportiveness, and Conscientiousness. DiSC is a system developed by the Carlson Learning Company, a Minneapolis-based firm that develops training tools for sales and business interactions.

D Is for Dominance

About 18 percent of the population brings a "D" attitude into your store or business. These people are direct and to the point. They tend to shape their environment by overcoming opposition and obstacles. Typical Dominance characteristics include the following:

- A desire for immediate results.

- The ability to make quick decisions.

- A tendency to question the status quo.

- A desire for direct answers.

- An environment relatively free of obstacles and controls.

- The ability to take charge of a situation.

How to win over D-style customers? Be brisk and efficient. Don't send them to another line to fill out a form; they want action right on the spot. If D-style customers don't feel that you're dealing with them effectively, they're apt to make a fuss loudly. You need to take swift action and appear to give the D's problem every bit of your attention. Stick to the facts: what happened, when, what specifically you can do to correct the situation. You don't want to offer an ultimatum, though. Because D's like to be in control, you want to offer several options for resolving the problem. But if you get results for them quickly, they're apt to reward you with loyalty.

I Is for Influence

Influence-style customers are talkers. This personality type, which accounts for about 28 percent of the population, likes people and needs to be around them. I-style customers are expressive, enthusiastic, and friendly. Other characteristics include the following:

- A desire to make a good impression.

- The ability to articulate a problem clearly.

- Enthusiasm.

- A desire to entertain.

- Optimism.

- Likes to be part of a group.

- A desire to be recognized.

- A tendency to dislike detail.

You need to give I-type customers time. They want to talk things over and build rapport with you; you need to do the same for them. If the D customer wants action, right now, the I customer wants to enjoy the process. Make lots of eye contact, hear the influence customer out, and engage him or her in talk. Respect I customers' sincerity, and show that you can be trusted to follow up on a situation. Be friendly and open, not tight-lipped. Ask for their opinions, and how they'd like to see a situation resolved. Chances are they'll appreciate the chance to be heard.

S Is for Supportiveness

One reason why people don't complain as much as they should is that most of them are S-type personalities—40 percent of the population, in fact. S-type people are accommodating and remarkably patient. They're willing to cooperate with others to ensure that a task is accomplished or problem solved. But that doesn't mean they're completely forgiving—they want solutions to work, and they want closure on their problem. S-type characteristics include the following:

- Predictable, consistent behavior.

- Patience.

- A desire to help others.

- Loyalty.

- Good listening skills.

- A desire to be appreciated.

- A dislike of conflict.

- Finding comfort in procedures.

When dealing with a S-type customer, stay calm. Be reassuring and sincere. Hear S customers out. Don't interrupt, and don't stick to the hard facts. Don't force them into a quick decision. Be casual and informal, and ask specific questions, such as *how* you can help fix the problem.

C Is for Conscientiousness

The C customer is a stickler for detail. Conscientiousness customers are a minority—14 percent of the world consists of C personality types. They're often the ones, though, who keep the world ordered and on time. They're often obsessed with neatness, and they want to know how things work. Characteristics include the following:

- Attention to detail.

- Analytical thinking.

- A tendency to weigh pros and cons.

- Diplomacy.
- Use of a systematic approach to deal with a problem.
- A desire for value and quality.
- A desire for control over as many factors as possible.
- Appreciation for a businesslike atmosphere.

Don't get emotional; C customers like facts and figures. Be prepared to back up anything you say to a C. Offer specific solutions to problems, but be willing to show flexibility. Be businesslike, not informal. C's don't care for too much chitchat. Be organized, and give time for a C person to come to a decision. Be patient—C-style customers like lots of time to think things over.

The Demographic Wild Card

The problem with people is that they're just so darned hard to classify. Not only are there different personality types; other factors can have an effect on what customers think about you and how you deal with them. The American Customer Satisfaction Index (ACSI), a survey conducted as a joint project of the University of Michigan, Arthur Andersen, and the American Society for Quality that covers the years 1994–1997, highlights three demographic areas that influence customer satisfaction.

Gender and age

On a 100-point scale, women rate their satisfaction as customers four points higher than men. This difference in ratings holds across all products and services and income levels. Why? One possibility is that women are simply better shoppers than men. They're generally better communicators than men, so they are often better able to express what they need. They're also often more willing to request—and therefore receive—assistance. On the other hand, it could mean that men are typically more involved with their purchases. They have higher standards, and they are more demanding.

In addition, the older consumers get, the more likely they are to list themselves as satisfied customers. Satisfaction rises uniformly with age from 18 years to 54, according the ACSI. And it really jumps above the age of 54. Why? Older customers have experienced different times: a Depression, several recessions, wars, and hardship. The younger the customer, the better the economic conditions that person is accustomed to. Younger consumers also are much more likely to expect continuous, rapid improvements in technology. Older shoppers may also have a wider range of experiences.

They know what they like, and they stick to their likes. In short, they're just mellower about their customer-service experiences.

Socioeconomic level

The ASCI divided consumers across three levels of combined income and education. The results? Wealthier, better-educated customers are generally markedly less satisfied with their customer experiences than those in lower brackets. Why? Because each customer group evaluates the perceived quality and value of their purchase within the range of what is affordable. A Lexus sedan may be a more expensive—and higher-quality—car than a Dodge minivan, but that fact does not affect the relative satisfaction the minivan customer has with the purchase.

The lesson here is clear. If you want to cater to more affluent customers, be prepared to earn their satisfaction. And don't be surprised if goods and services that most people would see as expensive and of high quality don't always pass muster. The more you make, the more you expect.

Geographic location

People in the Northeast are literally tough customers, according to the ASCI. They're the unhappiest customers nationwide. Customers in the West are the next least satisfied, according to the index. Those in the Midwest and the South tend to be most satisfied, which indicates that the regions' reputation for friendly service personnel is justified.

Root Cause Analysis

No one likes to get a complaint, says Fred Dublin. But a smart company will use complaints to build competitive advantage. Dublin is director of global logistics for AMP, Inc., a Pennsylvania company that is a global supplier of electrical parts and connectors.

Dublin has developed a system called "root cause analysis." Its seven steps help a company pinpoint problem areas and find solutions. While designed for a multimillion-dollar corporation, its systematic approach to fixing complaint sources could be emulated with advantage by any small business. "We've used this, and it works," Dublin says. "It helps you peel away the layers to find the true cause of a problem."

The seven steps are as follows:

1. **Mindset.** "Customer service depends on attitude," Dublin says. "You don't want to go into a fast-food restaurant and have some snarly sonofagun say, 'Whaddya want?'" So the first step is to make

customer service a core value. This starts at the top, and it must permeate the entire company culture.

2. **Develop cross-functional teams.** People have different perceptions of problems and complaints, Dublin says. So he advises that problem-solving teams consist of people with different backgrounds and company responsibilities. "You get to view the problem from a lot of different perspectives," he says.

3. **Develop input sources.** A company that wants to learn from complaints must work hard to get them, says Dublin. Merchandise returns, phone calls, comments made in a story—all can be sources. Also, when complaints come in, open case files on complaining customers using Microsoft Outlook or Microsoft Office 2000's Customer Manager, and flag the entries for follow-up. Pay close attention to what your customer data is telling you, and put this information in the hands of people who have the authority to act. Some companies report that field personnel tend to smooth over problems by buying a client lunch or handing them an on-the-spot discount, instead of reporting the problem back to the main office so it can be identified and fixed permanently.

4. **Perform a situation analysis.** Where are you now, and where do you want to be? That's the gap you have to work on, says Dublin. Ask yourself what it takes to win and keep business. When thinking about complaints, write down everything that might be a cause of the complaint. And consider your competitors—what are they doing to win your customers from you, and how might a customer's complaint about you affect that customer's decision to stay with you?

5. **Make a process map.** Make a chart that proposes several different solutions to a problem. "You have to resist saying, 'I think the answer is this,' and stopping there," says Dublin.

6. **Propose a solution.** Come up with ways to solve the problem. This must strike a balance between profitability and customer satisfaction.

7. **Implement.** Draw up a plan to fix your problem, and stick to it. Schedule frequent follow-up discussions to ensure that the problem is indeed being fixed. That will help you avoid what Dublin calls the "mandate and move on" syndrome, in which company managers simply declare a goal and then fail to supply the resources to see to it that the goal is reached.

The Angry Customer

Customers may have high expectations, but few believe any business or employee is perfect. It's how the company reacts to remedy a problem that leaves a lasting impression. Handled correctly, even a serious grievance can turn into a winning situation for your company. A customer who is angry or unhappy with your service and whom you can win over to your side isn't just a satisfied customer. He or she is a convert—and converts will spread the gospel of your customer service.

Your company policy should be to allow unhappy customers to tell you their gripes and give you a chance to demonstrate your determination to make them happy. The policy should spell out how, when, where, and by whom complaints or questions are handled. One person within the company should have the ultimate authority and responsibility for customer relations, although all employees should know your policy and how to implement it.

"Empowering" employees to solve problems is a popular theme. It can be an effective tool in resolving complaints. After all, people who already are unhappy are apt to go truly ballistic if they hear "Well, the manager isn't here right now" or (as the employee points to a line that stretches from here to Timbuktu) "You'll have to go stand in *that* line." Talk with employees about recent angry-customer incidents, how they were resolved, and how they should have been resolved.

You need to do more than simply fix problems, however. You should always be working to nip those problems in the bud. This requires that you set up a system for recording complaints. To do so, you need to take the following steps:

- Develop a recording method so that the date the problem occurred, a description of the problem, and any other pertinent information can be recorded.

- Investigate each problem and determine what caused it. You'll likely find that many problems represent the final breakdown in a customer's long-running difficulty.

- Determine a way to acknowledge that the customer is having a problem and that you are working to correct it. When answering a complaint, keep your emotions out of the response letter or phone call.

- Formulate solutions. Take into account both the technical and emotional issues involved. Important criteria to consider include warranty obligations, your customer's expectations, the cost/benefit ratio of alternative solutions, the public relations impact of a discontented customer, the fairness of your decision, and your ability to carry out the solution.

- Respond to the complaint. Your response should be clear, appropriate, and specific to the individual's complaint. Avoid form letters and technical jargon. Providing an explanation of your decision can preserve your customer's goodwill, even if he or she desired a different result.

- Follow up. Personally contact the customer following your response to verify that the matter has been resolved satisfactorily.

- Welcome the customer back, and make it clear that there are no hard feelings.

SAVE an Unhappy Customer

Jay Goltz, whose Artists' Frame Shop in Chicago is founded on the premise of great customer service, has developed an excellent method for dealing with unhappy customers. It's called SAVE—for Sympathize, Act, Vindicate, and Eat (I'll explain shortly).

First, sympathize

If someone is unhappy, few actions go as far as acknowledging that he or she has a reason to be angry, says Goltz. "Let customers know right off the bat that you understand why they're upset," says Goltz. "That can provide comfort to an unhappy customer." This is harder to do than it sounds, however. Employees who are facing an angry customer may feel under attack, and they may go into "attack" mode themselves. That's why so many complaints are rebuffed by an accusation—"Well, you didn't read the directions" or "You're the first person who's ever had a problem."

Second, act

Sympathy goes only so far, notes Goltz. Most unhappy customers want their problem resolved, not simply patted. Take action to show the customer that you intend to do all you can to solve the problem. Ideally, your employees should be trained to handle the majority of foreseeable problems. They should be trained to recognize likely problems, and they should be given a set of responses that you have authorized them to make.

Third, vindicate

When a customer at one of Goltz's stores finds an error in his or her order, he wants the customer to know that it's an unusual event—a step he calls *vindication*. He wants the customer to know that such a mistake is rare and not at all acceptable in one of his shops. "It's reassuring to tell a customer, 'An inspector checks every order. It's very unusual for them to miss something.'"

Last, eat something

Even if you fix the cause of a customer's dissatisfaction, says Goltz, you still are not even with that customer. "They didn't come to you to have a problem," he says. His advice is to "eat something." If the customer's meal is lousy, don't charge for it—*and* give the customer a coupon for $20 off on the next visit to your restaurant. If an order isn't ready when it's supposed to be, and if the customer has come to pick up the order, arrange to have it delivered. "Compensation has nothing to do with the actual cost of the problem," says Goltz. "It has to do with compensating them for the grief you've caused them."

The Fairness Factor

Much of what customers want when they're unhappy is plain, simple *fairness*. They what to believe they've been heard and that a company has made a sincere effort to fix the problem.

Yet many companies miss this test, says Dr. Stephen Brown, director the Center for Services Marketing and Management at the Arizona State University College of Business. They fix a problem based on a predetermined formula, without thinking of what's really fair to the customer. As a result, the majority of customers who have had a service problem do not feel they were dealt with fairly.

According to Brown, fairness has three components:

1. **A fair process.** Fairness begins with the polices and procedures of a company, Brown says. A company needs to assume responsibility for the failure, develop a creative way for the customer to have to the problem solved, and be flexible in its handling of

problems to allow for individual situations. Customers perceive a company to be unfair when the complaint-resolution process is long and complicated and when no one with a firm seems intent on taking responsibility to solve the problem. A key point: Most people like to have their complaint resolved by the first person they contact rather than being passed up some (to them) invisible chain of command.

2. **A fair interaction.** Fair interactions, says Brown, involve demonstrations of politeness, concern, and honesty, as well as an explanation for the reason behind the initial failure and a sincere effort to resolve the problem. In short: An apology and a little hustle go a long way. Still, this is a tricky area. Customers who finally feel motivated enough to complain directly are often quite angry. This in turn makes employees defensive rather than polite. The situation can escalate rapidly if the employee is not authorized to handle a problem on the spot. The solution is to empower employees to handle most problems, and provide training on how to deal with situations where anger might take over for reason.

3. **A fair outcome.** When people have been inconvenienced, they expect to be compensated. Typically this can take the form of a refund, a credit, a correction to charges, or a replacement. An apology or other acknowledgment that a mistake was made also figures into whether a customer believes he or she was handled fairly. But most companies miss two key points. One is that simply refunding $10 for a broken $10 item does not necessarily make the customer "whole." The customer had to take the time to return the item, plus it may have caused him or her annoyance or wasted time. People who are impressed by the company's handling of a problem, says Brown, are compensated beyond the price of the item involved. In these cases, a company has recognized that a defective product is only part of the problem. Also, many people appreciate being given several options for how they wish to be compensated. Some may want a refund, others a discount on an upgraded item. Be flexible.

See Table 11-1 for examples of customer reactions to fair and unfair responses to complaints.

Outcome, Procedural, and Interactional Fairness

Fair	Unfair
Outcome fairness	
"The waitress agreed there was a problem. She took the sandwiches back to the kitchen and had them replaced. We were also given free soft drinks."	"The restaurant's refusal to refund our money or make up for the inconvenience and cold food was inexcusable."
"The auto repair shop was very thorough with my complaint. One week later I received a coupon for a free oil change and an apology from the shop owner."	"The situation was never remedied. Once they had my money, they disappeared."
"Since I had to come back to the store for a third time, the manager not only gave me an exchange, but also apologized repeatedly and gave me a $25 store coupon."	"If I wanted a refund, I had to go back to the store the next day. It's a 20-minute drive. The refund was barely worth the trouble."
	"All I wanted was for the ticket agent to apologize for doubting my story. I never got the apology."
Procedural fairness	
"The hotel manager said that it didn't matter to her who was at fault, and that she would take responsibility for the problem immediately."	"They should have assisted me with the problem instead of giving me a phone number to call. No one returned my calls."
"The representative was pleasant and she quickly resolved the problem."	"Not only did the hotel ruin my vacation, but then I was accused of causing the problem!"
"The sales manager called me back one week after my complaint to check whether the problem was handled to my satisfaction."	"I had to tell my problem to too many people. I had to get irate in order to speak with the manager, who was apparently the only one who could fix things."

(continued)

Table 11-1

Comments adapted from a survey of customer satisfaction conducted by Stephen Brown of the Arizona State University School of Business.

Table 11-1 (*continued*)

Fair	Unfair
Interactional fairness	
"The loan officer was very courteous and knowledgeable—he kept me informed about the progress of the complaint."	"The person who handled my complaint about the faulty repair wasn't going to do anything about it and didn't seem to care."
"The manager had a good attitude. She wanted to make sure I left satisfied."	"The receptionist was very rude; she made it seem as though the doctor's time was important but mine was not."
"The teller explained that the delays were caused by a power outage that had occurred that morning. He went through a lot of files so that I would not have to come back the next day."	"They lied to me about the free Pepsi and they wouldn't give me a clear explanation of why the pizza was so late to begin with."

Follow Up

A final step, says Brown, is to keep track of complaining customers. Why? One reason is to ensure that they do indeed believe you have solved their problem. In many cases, the complaining customer, once satisfied, becomes one of your most loyal advocates. But if you blow it again, don't count on forgiveness.

One way to handle this is to put dissatisfied customers on a "VIP" list and ensure that they are given special attention when their order is processed or a sales call made. Use the customer contact features in Microsoft Outlook or Microsoft Office 2000's Customer Manager to start case files on customers who have had problems.

Such a system also helps protect a company in those very rare instances when customers abuse company policies on such matters as 100 percent unconditional guarantees. Some customers may seek benefits not normally provided by the company and may end up costing a company more than the customer is worth in profit. Some people may simply be crooks. By maintaining files on customers who have had a complaint, you can identify individuals who seem to be taking advantage of you. In those cases, it may be time for you to politely tell the customers that because you cannot meet their standards, they might be happier trying another company.

Chapter 12

The Pulse of the Customer

Of course, resolving complaints and conflicts effectively is one thing; working to prevent service issues from arising is another thing. Keeping an eye on service quality should be a daily part of your life. Here's a tasty role model for you: John Schnatter, founder of the fast-growing Papa John's pizza franchises. Schnatter regularly drops in unannounced at different restaurant units, and woe to the unprepared store. If the crust isn't quite right, or if the sauce is a little flat, there had better be answers. "Pizza is Schnatter's life, and he takes it very seriously," a colleague said about Schnatter to a writer for *Success* magazine.

With those visits, Schnatter is doing something every business owner should do: assessing quality as a customer would view it. Experts on customer service regularly recommend that a company owner spend time acting like a customer for his or her own product or service. Let's say your company offers a toll-free number for technical assistance. Have you called it lately? Have you ordered something from the Web site or over the phone? Tried to reach a particular office? Gone through the front door of the store rather than the back?

It's also vital to listen to your customers. Poor listening is a regular source of customer-service problems. Botched orders, missed credit-card numbers, a misheard telephone number—those little mistakes and more present customers with an endless barrage of small frustrations. Some companies actually train their employees in the art of listening. At Dick's Drive-Ins, a chain of back-to-the-fifties hamburger drive-ins in the Seattle area, employees pride themselves on memorizing the most complex orders without jotting them down or punching a photo-illustrated key on a cash register. To complete the act, they then announce the total due, all from mental addition. Starbucks, home of the "single-shot grande extra-hot vanilla cappuccino," uses a system for hearing and repeating orders developed solely to avoid errors.

Listening well is an inordinately powerful tool. People have an innate desire to be heard, a desire that has become almost touching in this day of constant noise. Think about it: When was the last time you spent any more than four or five minutes fully engaged in *listening* to someone? Probably it was a long time ago. Journalists frequently see the result of this listening deficit. As interviewers, they're paid to listen. And most will tell you that the hard part is not getting people to open up; it is getting them to shut up. Once the torrent of talk starts, it is hard to stop.

Learn to listen to customers. Try to find time just to talk with them, to see how things are going, if there are any aches and pains, what they think of your product or your employees. Try to chat with customers during quiet times in the store or at the end of a client meeting. Make a conscious effort to avoid turning a conversation with a customer or client into a verbal duel, where each remark he or she makes receives an immediate rejoinder from you. Remember that the vast majority of customers came to you because they have a problem to solve, whether it's getting a new pair of shoes because their old ones are worn out or refocusing their company's brand image. Find out what that problem is. Find out how you can solve it. Just try to *listen*.

Survey Your Way to Customer Understanding

For a business, listening must also include consciously and deliberately seeking out the input of customers. You want to get input on everything you can that relates to the customer's experience with you—the quality of the merchandise or services, the customer experience, and more. Service-smart companies are avid listeners, constantly seeking input from customers. Generally, companies can use these tools to gather information about themselves:

- Telephone surveys

- Mail surveys

- Mystery shoppers

- Focus groups

- Lost customer surveys

No one method is perfect for every company. Many larger companies, in fact, employ all of these techniques—and others. And each method brings its own kind of results. Terry Sutton, director of planning and business development for L.L. Bean, rates them on a scale ranging from highly *qualitative* to highly *quantitative*. Qualitative results are best gathered from individual or small-group interviews. This information tends to be highly impressionistic and anecdotal. It's used at L.L. Bean, for instance, to determine broad directions for product design and offerings. Written surveys, at the other end of the spectrum, are best used in large quantities to gather meaningful statistical data about such issues as customers' impressions with service handling or product quality. Telephone surveys fall somewhere in between.

What Do You Want to Learn?

Of course, the first step in developing an effective survey is to understand what you want to learn. First, survey yourself with questions such as these:

- What would I like to learn in a survey?

- What would make me more competitive?

- I want to change five things about my company, but I can afford to change only one right now. What would be most cost-effective?

- Am I doing better than I was a year ago?

- Am I doing worse?

- What will I or can I do with the information I might receive?

That last question is particularly important. If you aren't prepared to act on what a survey might tell you, don't waste your money and your customers' time.

Next turn your attention to what you'd like to learn. Potential survey subjects include the following:

- How helpful and knowledgeable your staff is

- Whether customers find your store layout and schedule useful

- How well your product works
- Whether your delivery system is adequate
- Whether your billing system is accurate and easy for clients to use
- Your standing versus competitors
- What customers like most and least about your company or service

Don't try to do too much in a survey. Focus each survey on one aspect of your business, seeking depth in a particular subject rather than breadth.

Once you have a grip on what you want to learn, it's time to decide on a survey technique. Let's look at the different approaches available.

Telephone surveys

Telephone surveys can give you quick feedback, but they can be difficult to do properly. For a small business, it's best to work from a "house list" of known customers; they're most apt to respond favorably to your call. For larger surveys, you'll need to hire a surveying firm that can develop a list of subjects that fit the demographic group you want to survey. Recipients of phone surveys tend to lump them together with other forms of telemarketing, making it difficult for the surveyor to win over the person being called. To make a telephone survey work optimally, follow these tips:

- Make the survey short. It is far better to get in-depth responses to a few questions than poor answers to a long list of questions.
- Make sure those doing the survey follow the basic script carefully. Seemingly minor variations can badly skew the data.
- Make the questions concrete and specific. Abstract questions may be difficult to answer. Be wary as well of rating systems that have too many steps, such as a 1 through 10 scale. A 1 through 5 scale will likely work better.

Mail surveys

Surveys sent by mail can be enclosed with an order or simply sent to a select list of customers. Again, for a small business, working from a house list is best. Mail surveys can supply a large amount of high-quality information. They're much less intrusive than a telephone survey, and once recipients decide to take the time to complete a survey, they're apt to give it a surprising amount of thought. Generally speaking, a mail survey is simply a direct-marketing piece, and the same rules apply as in writing a good direct-mail pitch. You want to assemble a package that is attractive and easy to read. You want to make it easy for the recipient to return the survey form to you. And you want to write a cover letter that briefly asks for (and thanks) the responder for his or her time. Tell why you're taking this survey and how you intend to use the results. Make the recipient believe he or she has a stake in future dealings with your business. You may also wish to add an incentive to completing the form, such as a discount coupon or a chance to win a gift in a drawing.

Questions in mail surveys typically model several common styles. These include the following:

Yes/no questions

Were you served promptly?   _Yes   _No

Was the project completed on time?   _Yes   _No

Did the results of our work meet your needs? _Yes   _No

Five-point scale (also called a Likert scale)

Our service is prompt.   _ Strongly agree. _ Agree. _ It was OK. _ Disagree. _Strongly disagree.

Our client representative understands your business.   _ Strongly agree. _ Agree. _ Somewhat agree. _ Disagree. _ Strongly disagree.

General questions about satisfaction

What is your satisfaction level with our service/product? _ Very high. _ High. _ So-so. _ Low. _ Very low.

Will you use our service or product again? _ Very likely. _ Likely. _ Perhaps. _ Unlikely. _ Very unlikely.

It's best to mix several types of questions in the survey. Other guidelines about mail surveys can help ensure success:

- Test the mailed survey to verify that it doesn't take more than three or four minutes to complete.

- Consider how the survey data will be entered into a database. Design the response form so that you, an employee, or a survey company can easily record the data. This approach will save money and reduce errors.

- Think carefully about the questions. Avoid leading questions ("Would you buy more from us if our prices were lower?") or those that show bias. And avoid ambiguous questions, such as "How would you rate the appearance of our store?" Appearance may mean different things to different people. Clean? Well designed? Conveniently located?

- Rather than simply dropping the survey in the mail, try to actively promote its use. Ask employees to distribute surveys to customers, and then ask customers if they have had a chance to complete the survey.

- Leave room for comments. Your most useful responses, particularly on surveys that go to a relatively small sample, may be anecdotal.

- Ask questions about the competition. See how your services or products compare.

Mystery shoppers

A popular tool, mystery shoppers can give real-world evaluations of how well your employees are giving customer service. The technique is obvious: You simply have someone visit one of your stores or outlets to see how your front-line service people deal with customers. You can use mystery shoppers to evaluate wait times, general friendliness and enthusiasm of employees, and the appearance of facilities. A mystery shopper can be almost anyone as long as he or she is not apt to be recognized by employees. It could be an employee from another store or department or a shopper hired by a firm that does this sort of survey work. Be sure, however, that the mystery shopper has clear instructions on what to evaluate.

Encourage your employees to be their own mystery shoppers by checking up on the competition. If you run a retail store, for instance, budget a small but meaningful sum of money that you distribute to employees so they can purchase an item for themselves—provided that they brief you on their experiences at a different store and offer suggestions for how to improve service in your own facility in exchange for the gift.

Focus groups

A focus group is a collection of 6 to 12 people (10 is probably ideal) that you bring together to discuss service and product quality. As L.L. Bean does, you can use a focus group to talk about different product or service offerings, evaluate different marketing approaches, or discuss pricing and features for your products.

To be effective, a focus group requires careful planning. You need to determine exactly what sort of feedback you want to receive and how you will receive it. Simply calling a group of people together to yak about your store won't work. Try to target regular customers who know your work or products well as potential participants. Also try to draw on as wide a demographic profile as possible. In your invitation, create the impression that the prospective participant has been selected because he or she meets certain criteria—he or she is a premium-level customer, for instance. Hey, you're playing a little bit to the customer's vanity. Explain the purpose of the focus group, the sorts of things the participants will be asked to do, and how long they can expect to be involved (an hour to 90 minutes is best).

Chances are, you'll need to offer an incentive. This can be in the form of a straight cash payment (usually $25 to $30), a gift certificate, or a choice of several products from your company or store line. Some focus group managers have found that a drawing for a larger prize is a strong draw. And offering soft drinks and food certainly helps.

Kathy Leck of The Lake Forest Graduate School of Management frequently runs focus groups for clients. One of the big challenges, she says, is ensuring that everyone in the focus group is given a chance to participate. In almost any group, she notes, one or two people are apt to quickly dominate the discussion (usually the "D" personality type mentioned in Chapter 11). To avoid that, she likes to structure focus groups along what she calls *nominal* or *structured* lines. In a nominal focus group, she places posters on the walls in the meeting room with broad topics written on each poster ("Pricing," "Store layout," "Employee knowledge"). Participants then use sticky notes to write down what things have enhanced or worked against their satisfaction with that particular issue. Participants then place their sticky notes on the appropriate posters, and the moderator uses the notes as a springboard for discussion, looking for common themes in the notes and consensus on a single key point in each category.

In a *structured* approach, group members are asked a question and then given a minute or so to write a response. The moderator then asks each individual to comment on his or her written remarks. When all have spoken, the group can then discuss the comments and reach a consensus on the most important points.

For each point covered in the focus group, says Leck, it's important for the moderator to recap and repeat what seems to be the group consensus. "What they said and what you think you heard might be two different things," she says. "So say to them, 'Now, the way I understand it, you think this.'"

Other key points to consider when holding a focus group include the following:

- Make sure the room is conducive to a productive meeting. It should be brightly lighted, free of distractions, and comfortable. Seat people in a circle or horseshoe shape so that they can see one another.

- Start on time, and hand out an agenda.

- Have a neutral moderator. Using a company employee may work if he or she is trained to moderate focus groups, but usually an "insider" finds it difficult to be neutral.

- Record the meetings with a video camera or tape recorder. Be sure that everyone knows the session is being recorded.

- Follow up by sending a thank-you note to all participants. Recap the meeting for them, and describe how you intend to use the information you gathered. "Once people have given their opinion, they want to feel they've been heard," says Leck. "But a lot of businesses ignore this after a focus group."

"Lost customer" surveys

Have you had a customer who seemed to be a loyal and frequent visitor, and then just up and vanished one day? It could well be that life circumstances have led that customer away—he or she moved or changed jobs so that the commute pattern is different. It could also be that the customer was angered or disappointed in your service and has simply taken his or her business elsewhere. Researching lost customers yields a high return for the time invested. If you can win back customers who have left you, they often become extremely loyal and valuable customers.

But as you might expect, conducting a lost-customer survey can be a little tricky. If the customer has indeed "left" you, he or she might be angry at you or a little embarrassed to be singled out for attention. For that reason, lost-customer surveys typically are conducted over the telephone. You want to assure the customers that you intend to take as little of their time as possible, but that you had noticed their absence and wanted to chat with them in the hope of winning them back or learning how you could better serve other customers in the future.

To conduct a lost customer survey properly, follow these steps:

- Decide what you mean by "lost." A customer who has not come in for three months? Six months? A year? Also, it's most profitable to pursue lost customers who were excellent customers or who held the promise of becoming one. To do this you need good customer records. Microsoft Customer Manager, for instance, allows you to search for customers in your database under a wide range of parameters, such as their annual expenditures and their frequency of (or last) visit.

- Use a neutral caller. Don't allow the sales manager for the lost account to make the call.

- Be prepared to probe the customer's responses, albeit carefully. If a customer says "I didn't care for your service," try to learn the exact nature of his or her unhappiness. It may be that you weren't doing anything "wrong" by your standards; the customer simply found service he or she liked better. If so, learn what the competitor is offering.

- Ask what you could do to win the customer's favor. Let the customer express a possible solution.

You've Got the Data; Now What?

Of course, even the best customer survey is worthless unless analyzed properly. This can be a tricky business; even the experts often disagree on the best method for sampling a group and analyzing the responses. And those untrained in statistical analysis may find that they've just steered their company in the wrong direction because of faulty data or incorrect analysis.

Common mistakes that crop up in small-business survey techniques include the following:

- Lack of validity caused by survey techniques that measure the wrong thing.

- Bias on the part of the questionnaire or observer. You must work hard to ensure that your questions and the survey design encourage neutrality.

- Answers that are biased because the survey was structured to be leading or seemed to encourage a "correct" response.

- Failure to understand the survey participants who do not respond. Let's say you send out 1000 surveys and 100 people respond. Among those respondents 50 people indicate they would like to see a particular new service. Your assumptions could take two extremes: In one, you assume that half of all those surveyed want to see the new service, or 50 percent of the sample pool. You also could assume that only those 50 people who say they want to see the new service really do, or 5 percent of the sample. Either assumption is likely to be incorrect. But you must develop a way to allow for those who did not respond and try to understand their real feelings.

- Sample errors cause many survey problems. If you are using a house list, do not survey only those people whose names you recognize. Do all you can to ensure that the survey sample is random and is as large as practical. The type of questions you ask also can influence the sample size. If you offer respondents a wide array of answer options (say, 10), then you require a larger number of responses than if you ask questions with 5 or 2 options (that is, yes/no) possible answers.

In any case, it's important that you or an employee evaluating the survey results understand basic statistical analysis. For that reason, many businesses contract out surveys to companies that do this sort of work regularly.

Still, for an entrepreneur with a good grasp of statistics, Microsoft Excel is a powerful tool for analyzing survey results. While it may seem a daunting task to enter data, even 1000 survey returns can be dealt with quickly if the survey was designed to allow for easy reading and data-entry (an important consideration in its design).

Chapter 13

The Client Connection

Georgia Patrick is a consultant who helps clients plan meetings and handle public relations. But that definition may be too narrow, she says. "I used to call myself a consultant," she says. "Now I say I'm in the customer business. If they're willing to pay me to learn something that will help them, then I'll do it."

Patrick, whose firm, The Communicators, is based in Maryland, is one of thousands of small business owners across the United States who regularly deal with one of the trickiest customer-service issues around: working with a relatively small number of demanding clients as a consultant and advisor. Attorneys, public relations experts, advertising agencies, engineering firms all share a similar challenge: the care and feeding of clients.

For Patrick, that means finding out what clients want and then doing all she can to fulfill their needs and expectations. Her goal: to make herself irreplaceable. "The more they invest in me, the harder it is for them to go somewhere else," she says. And as a consultant, you want to be in a position where your clients see you as invaluable. Not only will they reward you with more and bigger assignments, but you'll find yourself doing the work that led you to launch your consulting business rather than hustling for new jobs. A large company can afford to suffer some client "churn" as clients or customers move on to other companies or products. But a small consulting business cannot. "The cost of acquiring a client in consulting is just exorbitant," says Marilyn Hawkins, a public relations advisor in

Seattle. "The more you can keep the clients and business that you have, the more successful you can be."

Developing Your Own "Brand"

But how to build that strong relationship between you and your clients? Start by thinking of the service you provide as the "brand" by which the client recognizes and remembers you. "We spend a great deal of time with companies of all types talking to them about brand awareness and brand building," says Lee Duffey, founder and chief executive of Duffey Communications, Inc., an Atlanta-based public relations and advertising agency. "The big question that small business owners have to ask is 'What is my brand?'"

"A brand," Duffey says, "is a promise you make to your customers. Each time you work with a client, your 'brand' is an artifact of that interaction. Are you attentive? Are you listening? Customer service to your clients is one of the major foundations of brand building. So consider yourself as a brand, and always ask yourself: How can I fulfill that promise?"

Developing your own brand focused on your clients will ensure that you develop a lasting relationship with them. Even though all consultants' client work varies, Duffey and others say there are three essential elements to building and maintaining long-term, profitable relationships with clients:

Listening

Becoming a resource

Staying ahead of your clients

Learning to Listen

Hearing is the physical act your ears perform when they capture sound waves. *Listening* is what happens in your brain when those sound waves are converted to recognized speech. And by every indication, listening is a dying art. "Good listening takes time," Kathy Thompson, who teaches courses in conversation at Alverno College in Milwaukee, told the *Wall Street Journal*. "But we're always in a hurry. Mentally we're saying 'get to the point.'" Others blame radio or television for turning us into a nation of sound-bite listeners. Or there's the go-go twentieth-century American culture, in which listening is seen as passive and talking is seen as dynamic—and lucrative. Has anyone heard of a "listen show"? Hardly. Everybody wants to yak.

Consultants face a particular challenge when it comes to teaching themselves to listen. "Any professional, whether a doctor or accountant or engineer or lawyer, is taught that they're a smart professional and that they need to provide value to a client by constantly telling them what to do," says Mary Jane Pioli, a business coach who has worked with major legal and accounting firms as a marketing director. "They think their clients are paying us to tell them what we think about things. But consultants lose clients because they don't *listen* to their clients. It's a really powerful tool."

Pioli uses what she calls an "85/15" scale to assess how well she or the professionals with whom she works are listening to a client. "If the client is talking about 85 percent of the time, and you're talking 15 percent of the time, then that's about right," she says. "If it's the other way around, then the relationship is not doing too well."

The fact is that listening well is not a passive exercise. Our impulses to butt in or think about that golf shot we fluffed on the seventh hole yesterday are so strong that it takes a conscious, willed effort to pay full attention to someone else for anything more than 5 or 10 minutes. "I call it 'generous listening,'" says Katharine Halpin, a Phoenix-based CPA and business-practices coach. "It's not just what is said, but also noticing what is unsaid."

How can you develop those skills that lead to "generous listening"? Start here:

- Ask questions. What are the client's objectives for the project? What led the client to hire a consultant? Why were you hired? How does the company want to measure results? Companies hire outside help to solve specific problems, take the slack left by downsizing, or bring in expertise the company can't afford to hire full time. Find out what led this client to you.

- When talking with a client, try to ask questions that react to a comment he or she just made. That way you'll dig deeper into what the client is trying to express. Not sure what to ask? Here are three sure-fire, all-purpose questions that will get you more information:

  Why do you say that?

  Can you give me an example?

  I'm sorry, I don't quite follow. What do you mean by that?

- Don't make a conversation with a client a game of verbal Ping-Pong. Studies of human interaction show that most people listen to what is being said to them solely so they can figure out something to say back. In other words, they're not "listening" at all—they're simply waiting for their turn to talk.

- Try to ask "open" questions. A question such as "What's your budget for this project?" is a "closed" question. Only one answer is possible. A question such as "How did you develop the budget for this project?" will give you insight into the client's thought process.

- Show interest. In face-to-face meetings, remember that as much as 90 percent of communication is nonverbal. Focus on the person who is talking, and reinforce his or her words with small nonverbal encouragements such as head nods. Semi-verbal "response triggers" such as simply saying "uh-huh" often are surprisingly effective at keeping people talking. And don't feel obligated to leap in with a pithy remark when a client seems to have finished a remark. A little silence goes a long way toward eliciting further words.

- Focus. It's rare today to pay full attention to anything. Why? In part because people tend to talk at about 50 words per minute. But we're capable of listening at 500 words per minute, meaning there can be a lot of time for daydreaming. When the client is doing the talking, though, you must pay close attention.

Listening is the first step in developing strong communications with your client. When you're talking with a client early in the relationship, identify the client's communication style. Much communication today is not face-to-face, but listening to what is being communicated remains extremely important. Does a client like voice mail? E-mail? Fax? Ask—he or she will tell you.

Once you know how your customer prefers to receive information, put the client in your communications loop. Keep the client informed on the status of your project. Ask how often he or she wants to be appraised of progress. Weekly? Monthly? "Even if nothing is happening, drop the client a note," says Pioli. "You want them to know that you're thinking about them."

Top 10 Client Expectations

Marilyn Hawkins, founder and principal of Hawkins & Co. Public Relations, Inc., borrows a little from David Letterman in developing her own list of the Top 10 Things Clients Want from Their Consultants. No matter what your line of consulting work, your clients will appreciate you all the more if you keep these traits of "ideal" consultants in mind:

Respond quickly. They return phone calls, e-mail messages, fax requests, and written inquiries promptly and thoroughly.

Set and meet deadlines. Their internal clock is set to the *client's* time, not solely to their agency's time.

Offer truly tailored recommendations. Off-the-shelf or half-baked solutions just won't do.

Develop creative—but not goofy—strategies and tactics. Anyone can come up with average ideas; clients are looking for that extra edge they know they don't have in-house.

Be results-oriented, not merely a planning, process, or billable-hours freak.

Have in-depth understanding of the issues the client is facing. They bring insight and skills to the party, rather than learning on the job.

Provide realistic criteria for evaluation. They don't wait to be asked "How will we know if this works?"

Be conscious of the role of personal chemistry. People like people like themselves. To gain their clients' trust, sometimes consultants must practice the art of "mirroring."

Show true enthusiasm about any assignment. Enthusiasm is contagious (and so is disinterest and ennui).

Teach important concepts and skills. But never in a condescending fashion.

Be a Resource

As Lee Duffey says, a client hires you because he or she needs capabilities he or she lacks. "We're adding arms and legs for the client," he says. "We want to add to what the company knows." This includes bringing an outside perspective to a company that may have become a creature of habit or whose executives have been inside the firm so long that their focus has narrowed. "It's always interesting to speak to a client that says 'we're flawless,' then talk to their competitors or peers and find that isn't quite the case."

To be a resource, however, means you must understand both the client's *business* and his or her *industry*. Understanding the client's business comes from asking questions and listening to the answers. Understanding the industry comes from tapping into the information stream that comes out of that industry—whether it is telecommunications, the health industry, or fast food. What magazines about the business are sitting in the client's reception area? Subscribe to them. Attend trade shows that a client might attend. Join the same professional groups as the client, and attend his or her dinner programs and other events.

This is one more area where technology has made it much easier for a consultant. Databases, Web-based publications, and Internet sites help a consultant tap into a stream of intelligence about a client's industry. Once you have that information, use it. Chances are, few clients really keep up with news in their own industry. But as a consultant you should make it your *business* to do so. Duffey's company, for instance, keeps an eye on Profnet, a national Web-based query system used by many journalists seeking story leads, and puts clients in touch with those reporters. Duffey also has set up search robots that watch Web newsgroups, providing an early warning should a controversial issue affecting a Duffey client emerge. Duffey Communications also regularly conducts focus groups with client customers and conducts mail and phone surveys.

Being a resource also means that you can predict what the client needs before the client knows he or she will need it. Georgia Patrick uses contact management software to keep a detailed file on every one of her clients. She opens a file whenever a client calls, developing a record of that client's needs and habits. "I know when every organization I work with is going to meet," she says, "and whether they plan ahead for these things or plan in a frenzy at the last minute. So I'm able to anticipate things. I'll call 'em up and ask 'So, are you doing a PowerPoint presentation this year?' and they'll say, 'Georgia—how did you know? We've gotta get this thing going and don't know where to start.'"

"Instead of my saying, 'Whaddya want?' I'm able to remember. People have habits, and they do the same stuff pretty much the same time every year."

Using technology to track and remind you of your customer's needs leads to the relationship building that can help you keep today's finicky customers. But, says Kathleen Allen, a business professor at the University of Southern California, "I don't see enough businesses making good use of contact managers such as a database or something more sophisticated. That way, at the very time you're talking to someone you can be entering information about them—what they need, what they're doing next. That way, you can know things about a customer that a typical company doesn't know. And everyone agrees that the most powerful competitive advantage you can have is information about your customer."

Patrick's clients certainly sense this. "I may not deal with Georgia for a year or so, but we'll start talking about a project and it's evident she's been tracking our organization," says Charles Van Horn, executive director of the trade group ITA, which represents manufacturers of recording media. "She'll quote things to me that she's read in our association publication." When Patrick was recently prepping a speaker for an ITA meeting, for instance, she plied the speaker with so much data on the group that Van Horn recalls the speaker as the best prepared he's seen. "Even when I work with a speaker's bureau, they don't do as good a job of preparing the speaker," he says.

Much of the task of keeping ahead of clients comes from truly understanding what motivates them. "Remember, they're human beings," says Marilyn Hawkins. "Are they motivated by a desire to succeed? Are they 'win' oriented? Or do they just hate to lose? Are they short-timers, looking for a quick solution, or do they see what they're working on right now as a stop along a longer road? When you're working with a client, you have to see these factors."

Put the Client First

As a consultant, your main task is to help your client solve a problem. "Nobody hires a consultant as a hobby," says Duffey. "There's always a business reason." And as a consultant you must always remember that a client will be impressed only by what works for him or her. That sounds self-evident, but there are several ways in which consultants can inadvertently do work that puts themselves first, not the client.

One way to make that mistake is to get caught up in your own creativity. "We're not rocket scientists—we are craftspeople who refine common sense," says Lee Duffey. "If you come up with a proposal that is so outlandish and far off center that it gets noticed for that reason alone, it won't be adopted by the client." In other words, don't worry about how your work will look in the awards contest for the local chapter. Worry about whether what you do will help your client.

Moreover, there are times when your client may not want your best possible work, either for deadline reasons or budget reasons. You need to be able to listen to the client and hear the message when he or she says that just OK will do fine. "There are times when a client just wants a 'C' solution, and the consultant wants to do an 'A' solution—and bill them for it," says Mary Jane Pioli. So talk to the client about his or her needs and expectations for a particular job.

Also, as a principal in your company you need to know when and how to get directly involved in a project and when to let your own associates handle a client's needs. That takes careful thought and is an assessment that your familiarity with the client and his or her needs supports. There are times when you don't need to be present, when your associates are quite capable of taking care of the work that needs to be done. At key points in the assignment, though, it's imperative that the principal weigh in, sometimes on a billable basis, sometimes not. "Sometimes you just want to show up on a nonbillable basis and let the client know that you're in control and checking on what your people are doing," says Pioli.

Last, be patient. Developing a strong working relationship with a client takes time. "Clients won't love you until they trust you," says Marilyn Hawkins. "And they won't trust you until they believe in you."

Nice Touches

Maintaining good client relationships also means paying attention to the little things. These include the following:

- Celebrating a little with the client when a key project is finished. "I don't mean throw a big party," says Pioli, "I mean congratulate the client, send flowers or something, let them know it's been great working with them."

- Letting the client know you're thinking of them. Try to learn when they have key dates pending—a child's graduation or other event—and acknowledge it. Don't go overboard, of course, but try to give your business relationship a human touch. Says Lee Duffey, "What do you as an individual enjoy? You like to have someone call you and ask how it's going, or send you a card on your birthday, or you want a note that contains a news clip or an upcoming seminar that can enhance you professionally. If someone treats you like you're memorable, you'll have a stronger affinity for that person."

- Bending your business practices to your client, not the other way around. Find out how the client likes to be billed, on what cycle, and in what form.

- Taking frequent "customer surveys." Sit down with the client on a regular basis to ask how you're doing. "Ask them what you can do better," says Pioli. "And, ask them what you're doing *right*. A lot of times a client will think they need to talk only about what needs improvement, but you can learn just as much about what you're doing right."

Using Technology to Keep Clients "Top of Mind"

The key, says Jon Ferrara, is to make sure you're in the customer's brain at the time he or she makes the critical buying decision. "That's keeping the customer 'top of mind,'" says Ferrara, executive vice president of GoldMine Software Corp., makers of GoldMine contact-management software. But that's easier said than done. A study by the Harvard Business School demonstrated that it may take as many as six or seven customer "touches"—occasions when your product or service is brought to the attention of a customer—before that person is really ready to buy. Yet often a company believes that one or two touches are adequate—usually, a brochure and a few phone calls.

Ferrara outlines a scenario in which today's sophisticated contact managers can help provide superior customer management. Let's say you have a client that appears in trade shows, and your business helps set up the booths. "You could put that customer in your contact manager and six months before the show start touching them," says Ferrara. "You can say 'here's a checklist, and we can help you get ready.' Then at five months, you tell them what things should be done. And if they're behind, you have a coordinator who can assist you."

GoldMine, the product Ferrara's company has developed, is one of several powerful after-market contact management tools. It integrates well with Microsoft Office 2000, giving you exceptional control over your customer communication. But Office 2000 Small Business, Professional, and Premium versions also have an outstanding contact tool: Microsoft Customer Manager. You can make it your customer service assistant by customizing it to do the following:

*   Keep track of your contact's business information, including titles and responsibilities of people in their corporate orbit.

*   Build lists of "Hot Reports" that include such categories as Top Customers and Top Products. At a glance you can learn who your current sales leaders are.

*   Remind you of follow-up actions you need to take, such as making sales calls or mailing new-product information.

*   Coordinate meetings between you, sales personnel, and a customer's staff.

*   Keep track of past activities with a client so that when you call or write to the client, you have a complete record of past conversations or correspondences with him or her.

Using Outlook and Customer Manager to track contacts

Microsoft Outlook and Microsoft Small Business Customer Manager can simplify the process of tracking clients and customes while giving you powerful new tools for ensuring your customers get the attention they deserve.

First, ensure that you have Customer Manager properly installed. Click the Start button on the Windows taskbar, point to Programs, and point to Microsoft Office Small Business Tools. Customer Manager will be located in that folder. If it isn't, insert the Office 2000 CD, run the setup program, and install the tools.

Customer Manager gives you the option of important accounting information from an existing database or contact information from Outlook or other contact sources. You can also view the sample database included with Customer Manager to learn the ropes.

If you're opening an existing database, you may need to type your user name and password and press Enter to open the database. Once open, Customer Manager displays information in rows and fields (see Figure 13-1).

With the database open, get a feel for Customer Manager and the advanced contact management features in Microsoft Outlook 2000 by performing several tasks:

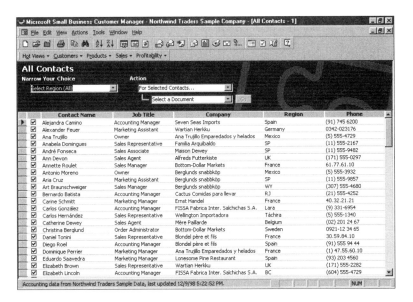

Figure 13-1

Contact information displayed in Customer Manager.

- Use Microsoft Word or Microsoft Publisher and Customer Manager to write a letter to a specific set of customers. To do so, first create a Hot View in Customer Manager that categorizes customers or employees by sales or other criteria. On the View menu, point to Go, and then click the Hot View that you want. You can categorize the database based on a wide range of criteria, including the following:

    Top customers

    Top salespeople

    Top products

    Customers by sales volume

    Customers by order volume

    Customers by product ordered

    Sales by customer

    Sales by region

    Profitability of product

    Profitability of customer

    Then click to deselect the check box to the left of any customer record that you want to exclude from the mailing. On the

Actions menu, point to For Selected Contacts, and then click the type of document you want to create. You'll then be offered a range of document templates to use (see Figure 13-2). Customer Manager includes more than 20 templates that can be easily customized to give customer contact materials a one-of-a-kind look.

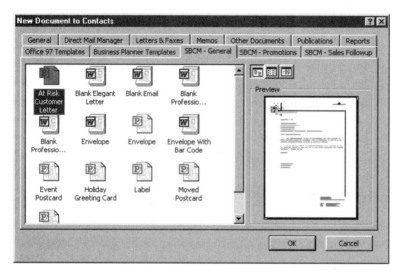

Figure 13-2

Selection of templates in Customer Manager.

- Track customer activities. In Customer Manager, click Options on the Tools menu. Then select whether to track Outlook events such as e-mail messages and Customer Manager events such as Word mailings, and then click OK. To review contact activity, select a specific customer record by clicking the row selector to the left of the contact name and check box. A right-pointing arrow will appear in the row selector. Then point to For Current Contact on the Actions menu. Click Show Contact Activity to see all recorded activity with that contact.

- Automate various tasks by taking advantage of the integration in Office 2000. For instance, you can select one or more customers from a customer list, and then select an automated task from the second Action drop-down list, such as sending a fax, an e-mail message, or another communication. All of that contact information—e-mail address, fax number, street address—follows that action right through to the finished document (see Figure 13-3).

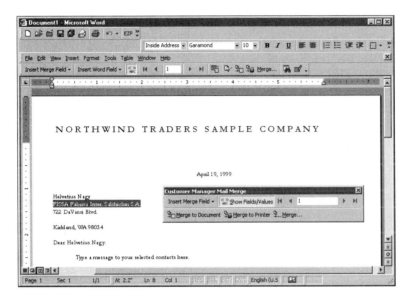

Figure 13-3
Use Office 2000 to automate "customized" communication with your clients.

Outlook's customer contact features

Outlook 2000 is another exceptional tool for managing customer relationships. In Outlook, you'll find new features that do the following:

- Make the program easier to use and manage

- Increase the usefulness of e-mail

- Improve collaboration over the Internet

- Improve remote user access

Using Outlook, you also can send e-mail messages or letters to contacts, link calendar appointments with one or many contacts, record contact discussions, and keep a history of all activity with a particular contact. Some useful capabilities include:

- **The ability to organize your contacts into folders.** Start Outlook 2000 and click the Contacts shortcut. Next click the Organize button on the toolbar. You can then group contacts by category (such as Suppliers or Key Contacts) or folder (such as Business or Personal).

- **Creating and documenting contact actitivies.** Simply select a contact listing in the Contacts folder, and then right-click that contact to show a shortcut menu of options—writing a letter, sending a message, placing a telephone call, arranging a meeting. You also can click New Journal Entry For Contact, which opens a Journal Entry form that allows you to take down notes from the conversation and time the length of a conversation for billing purposes (see Figure 13-4).

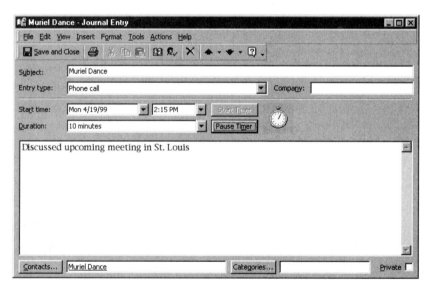

Figure 13-4

A Journal Entry form in Microsoft Outlook allows you to detail contacts with clients.

- **Automated record-keeping.** Outlook keeps track of all e-mail messages, phone calls, and other activities made using Outlook. Double-click a contact in the Contacts folder to open the contact listing. Click the Activities tab. Click the Show drop-down arrow, and then click the type of contact items to list. Outlook will list all activities with that client or simply those specified, such as e-mail messages, telephone calls, or tasks.

- **Links to contacts.** Using Outlook, you can link documents created in Word, such as a sales letter or business proposal, to the appropriate contact. Simply select a contact, point to Links on the Actions menu, and then click the type of item to link.

- **Mail merge to contacts.** Using Outlook, you can send personalized letters and e-mail messages and create labels and envelopes for mass mailings.

With the right techology and the right approach to handling clients, you can help ensure that you stay on top of that client's list of valued consultants and advisors.

Chapter 14

The Web and Your Customer

Customer service has probably known no greater revolution than that of the last half of the 1990s: the rise of the Internet. It's stunning, really. Customers can now seek out you—or your competitors—24 hours a day, 7 days a week, regardless of their location. Indeed, the Internet represents an amazing opportunity for a company that has a vision of excellent customer service and the ability to execute that vision in the Web environment.

A Web of Possibilities

For the Web-savvy company, the Internet opens up a host of opportunities in the customer-service arena. Forrester Research, Inc., for instance, surveyed financial institutions and found that answering a query by using the Internet costs a mere 4 cents per customer, versus $1.44 for a telephone call. Other advantages of Web-based customer service include the following:

- Around-the-clock access to your goods and services

- The ability to quickly and easily survey clients to gauge their satisfaction with your service and products

- The creation of informal electronic networks among your clients, who can help each other with questions and use your "office" as a place to meet, thereby strengthening your bond with them

- The ability to engage in high-tech, high-touch marketing that is quick and efficient yet delivers a level of laserlike customer targeting never before possible

- Deeper customer service by allowing the storage of data in quantity and of a type that would be impractical to do on paper

With the Web, the Bar Gets Higher

Endless possibilities, and equally vast risk. In Chapter 9 we talked about the "butterfly customer," who feels little loyalty because changes in shopping habits and greater mobility mean that he or she is not apt to bump into the local clothing-store owner in the coffee shop down the street. Well, now that customer may never even talk with you. He or she might simply flit through your Web site and decide to buy or not to buy from you within seconds before you have a chance to say "hello."

As a result, customer service may be even more crucial on the Internet than it is in the bricks-and-mortar world. Jeff Bezos, founder of leading Web marketer Amazon.com, has said that in the "old" retail world, 30 percent of a company's resources should go to customer service, 70 percent to marketing. But that figure is reversed for the Web, Bezos says—fully 70 percent of a company's efforts should be focused on creating a great experience for customers.

Why? Because in the Web world, the buyer holds most of the cards. Price comparison to a degree unimagined before now is possible, as well as easy access to a host of different companies or retailers. "The fundamental, earth-shaking event that's being shaped by e-business is a radical shift from buyers to sellers," says Andrew Brooks, CEO of Furniture.com, a Web-based furniture store renowned for its customer service. "What does that mean? Service is everything."

Thus far, though, most indications are that many Internet sites adhere to the notion that service is something. Just what, they're not sure. In this day where speed is seen as the consumer's newest entitlement—and the desire for speed has been accelerated by the Web—a survey by New York–based Jupiter Communications revealed this rather amazing statistic: 42 percent of the top-ranked Web sites surveyed took longer than five days to respond to a customer's e-mail query, never replied, or were not even accessible by e-mail. At the same time that survey was taken (November 1998), another Jupiter survey was in the works. It showed that overall customer satisfaction with online shopping was lower than it had been six months before, a troubling trend in a year when e-commerce was seen as about to make a breakthrough.

Common complaints about Web-based service? These are among the most widespread:

- **Failure to respond.** As the Jupiter report mentioned, many Web sites display an e-mail address, and then act startled when people use it. E-mail is seen by most of its users as an extremely quick communications technique. They rationally understand that a company may require 24 hours to respond, but in their heart of hearts they want an answer in an hour or two. Making customers wait for days creates an even larger gap between hope and reality.

- **Poor shipment and returns service.** Shipping and returns become a "back-end" problem when a company isn't set up to act like a real retail store. In a recent *New York Times* article, for instance, a customer told of ordering "Out of Sight" (a wonderful movie, by the way) on DVD from an online entertainment retailer. After several weeks, the movie had not arrived. The customer sent several e-mail messages to the retailer, with no response. Finally, in January 1999, three weeks later, the DVD appeared. At the same time, customers complained about fouled-up returns and credits that are too low.

- **No alternative contact.** Many customers will use a Web site simply to find you and your telephone number. It's true that you may be using the Web as a *replacement* for telephone contact, but that can be a mistake. The customer with the fouled-up DVD order, for instance, cited his inability to call the company on the telephone as major gripe. "From now on," he told the *Times,* "I'm not doing business with anybody who doesn't have a phone number."

- **Poor site design.** A book in itself. Customers want a site that is easy to understand and lets them get where they want to go with minimal button pushing. Many companies don't even test their front end, placing sites on the Web that literally don't work.

Understanding the Web Customer

Who is your Web visitor? Not long ago you could assume it was someone who was fairly technically savvy and who understood that in the world of technology (and in particular, the computer), there were bound to be glitches. That's changed. The year 1999 in particular saw a stampede of Web-site users who were new to the Internet. They're far less tolerant of mistakes and goof-ups and need a high level of customer service to assuage their nervousness about shopping or seeking services on the Internet.

That trend will continue, says Ken Allard, an analyst with Jupiter Communications. By 2002, he says, it is estimated that almost 20 million Web users will be over 50, with another 21 million between the ages of 2 and 12. "That could mean 41 million people online who aren't all that computer-savvy," he says. The implication: Web developers had better figure out how to create sites that are easy to use and that offer excellent customer service.

Things the Web Can Do

Remember—in many ways the Internet represents an entirely new way of doing business. It offers several customer-service methods that simply weren't available until recently.

E-mail

Many companies, says V. A. Shiva, still don't know what e-mail really is. Shiva is in a position to know—he created one of the world's first e-mail systems back in 1979 when he was 15 years old (connecting to Rutgers University) and went on to do advanced research at the Massachusetts Institute of Technology. Today, Shiva's company, General Interactive, Inc., is actively involved in helping companies enhance customer service and marketing over the Internet. Clients include JC Penney, Nike, and Calvin Klein Cosmetics. Its flagship product, EchoMail, is designed to help companies deal with the flood of e-mail messages that many now receive.

And just what is e-mail? "It's the most cost-effective, flexible, targeted communications system on the planet today," he says. "There's nothing else like it." Other advantages: It's fast, and unlike the telephone it is "asynchronous"—you don't have to be on the other end of the line to receive the message. But many companies tend to treat e-mail messages as another form of telephone calls, folding them into call-center operations, he says. That's a mistake—companies that are serious about using e-mail as a customer-service tool should establish an independent e-mail center to handle incoming messages.

Treated as a one-of-a-kind tool, e-mail can be an extremely powerful device for both marketing and customer service, Shiva says. Using

EchoMail, for instance, clients staffing a trade-show booth can scan a visitor's card and have that visitor's e-mail address instantly loaded into a server, with targeted marketing information quickly heading the booth visitor's way. Or, when highly technical e-mail queries come into an e-mail receiving center, they can quickly be routed to those company representatives most capable of answering the questions. That approach allows for much more flexible site management and employee scheduling than a traditional call center.

E-mail has other obvious advantages to the small business. It generates little or no paperwork. It cuts long-distance costs and saves on postage and letterhead. And it transcends time, allowing the recipient to answer at his or her own convenience. Making the most of e-mail, though, requires a conscious approach to handling it. This should include the following:

- Setting up a Web site so customers can route e-mail to the appropriate part of the company—whether sales, technical information, warranty support, or other matters.

- Developing a way to prioritize e-mail as it comes in. Unless the volume is completely overwhelming, one or several employees should be designated to scan incoming messages, mark them as urgent or not urgent, and then route them to the right department or individual.

Technology is fast making high-volume e-mail management possible. EchoMail, for instance, uses advanced pattern recognition and classification technology to scan incoming e-mail messages and route them to the correct service desk in a company or automatically respond. EchoMail also tracks e-mail so that a customer-service agent has on his or her computer screen a full record of communication with a customer. Nike, for instance, uses EchoMail to scan the thousands of messages it gets each day. Many queries—regarding merchandise or Michael Jordan—are answered automatically. EchoMail also can detect patterns in incoming messages, such as might occur if a company's public actions or a new product release results in angry customers.

A competitor of General Interactive is Mustang Software, whose Internet Message Center also handles routing and auto-response functions. One customer of Internet Message Center is Wall Street on Demand, a research service company that sells financial information under private labels such as Charles Schwab and Fidelity. After switching to Web-based information delivery from a phone- and fax-based system, Wall Street on Demand was swamped with e-mail messages. Installing Internet Message Center helped the company control the deluge. Now Internet Message Center helps Wall Street on Demand by doing the following:

- Tracking e-mail messages and generating reports on when the messages have been answered

- Indexing messages with an individual tracking number, making it easier to track the thread of a conversation that contains several messages

- Automatically distributing e-mail to call center agents at the company

- Automatically sending a customized "message received" message that gives the customer a tracking number and an estimated time before the full response is sent

Still, an elaborate add-on isn't always necessary, even for fairly busy systems. Recreational Equipment, Inc., for instance, maintains a site that receives about 1 million visitors a month who are looking for camping, climbing, or bicycling gear. Site administrator Matt Hyde, REI's vice president for online services, has configured Microsoft Outlook to filter incoming e-mail messages based on several keywords.

Watch that message!

Select e-mail responders who have a knack for written language. Just as a telephone call eliminates visual feedback that tells us whether the person with whom we are speaking is listening, e-mail takes away those voice inflections that say so much. Expressions used kiddingly over the telephone may sound cold and even intimidating in e-mail. Test your front-line e-mail people on this; have them respond to typical questions just as they would if the questions came in via e-mail. Don't assume that because they know the answer, they can put it in writing, clearly and graciously.

In general, be sure that your e-mail responses do the following:

- Contain subject lines that are descriptive and helpful. A message with no subject line may be winnowed by an e-mail filter or ignored by the recipient.

- Are formatted to be accepted by the widest possible array of Internet systems. Attachments in particular can be scrambled in transmission.

- Are succinct and to the point. One of e-mail's magic qualities is its ability to convey a great deal of information briefly. Good responses to customer queries are friendly in tone, but short.

- Repeat in abbreviated form the writer's letter or contain a copy of it. In the event the recipient doesn't read your *prompt* reply immediately, it's helpful to remind him or her about the request.

- Include a signature with the name of the person responding, his or her title, and as much contact information as you care to provide.

To auto-respond or not?

Auto-responders, also called "mailbots," are gadgets that monitor incoming e-mail. When a message arrives, the mailbot automatically fires a message back. Auto-responders are particularly useful on e-commerce sites, where they send to a shopper an acknowledgment that an order was placed and an order number or order recap. Auto-responders for more general purposes are becoming increasingly sophisticated, and many now look for keywords or phrases and try to determine what the message is about and tailor a response.

Still, several Web-savvy site managers believe auto-responders aren't yet ready to take over for more sophisticated mail filters: people. "They're pretty dang popular, but I wouldn't touch one with a 10-foot pole," says REI's Matt Hyde. "My view is that if you can generate an answer with an auto-responder, then there's something wrong with your site. People should be able to find the information there without having to send an e-mail to you that then generates another e-mail."

The Virtual Community

A second benefit of an Internet site is its ability to create a community of customers and people interested in your products and services. This is an incredibly powerful marketing tool. In marketing, says Michael Ponder, a Web consultant working with retailer JCPenney, there typically have been three steps: First, a company tries to get attention with advertising. Then it attempts to convince a customer that he or she has a need for the product. Finally, it supplies information to allow a customer to make a decision. If you have created a virtual community surrounding your products and Web site, you simply do not need to take those first two steps. You already have a crowd of interested people milling around your electronic lobby. You have their attention; they have a need. All you need to do is supply them with the appropriate information as new products or services become available.

There are two ways to go about creating this highly desirable situation. In one, you target a group of customers and work to open a conversation with them. For instance, Ponder's recent work with JCPenney had him developing a site for prospective or recent mothers. "That's a time of life with specialized needs," he says. "There's maternity clothes, play sets, and nursery furniture. There's also paint and carpeting and toys and local diaper service and formula and prenatal information." One possibility: Form partnerships with companies supplying products the site sponsor does not. With a prequalified audience of new parents or parents-to-be, such a site would be constantly relevant to those who visit it.

A well-designed Web site has the second benefit of creating a community of visitors who can help each other solve problems and raise issues that otherwise would take expensive staff support or extensive surveys. How? With a site newsgroup. Trellix Corp., developer of Trellix, a desktop productivity application designed to simplify the creation of Web pages, designed a newsgroup that gives customers a place to talk about the product and swap ideas. Visitors can talk to other customers, click buttons for e-mail support, and find white papers that discuss Trellix. "You can learn about the product on the Web site," says Paul Simpson, director of customer service and support for Trellix. "You don't have to talk to someone if you don't want to, or go to the Web site and see if you can figure things out on your own." About 40 percent of Trellix's customers use the site, Simpson says.

Simpson says such a Web-based system has some tremendous advantages over traditional telephone-based support. "If I talk to someone on the telephone and answer his or her question, five minutes later it evaporates," he says. "Now, I can archive a chat or an e-mail question and answer so that everyone has access to it."

One key to the success of such a site is to keep the content fresh. Trellix posts special bulletins when certain topics start popping up frequently from user queries. It also posts tips on using Trellix that are sent to the site daily and that can be sent to individual customers on a daily or weekly basis. "That just keeps growing," says Simpson. "I'd say 60 or 70 percent of our customers have signed up for our tips."

It's also a superb idea to be generous with your site, providing information as a service to visitors and not simply as an inducement to buy. One of the very best sites at this task is *Toysmart.com*, which until early 1999 was better known as Holt Educational Outlet. Visitors to *Toysmart.com* can find articles on topics ranging from breastfeeding to car seats to toy safety to good toys for traveling.

The Personal Touch

In a supreme irony, the Internet allows many e-commerce firms to regain the sort of one-on-one connection to its customers that was lost during the 1950s and 1960s with the rise of mass marketing and the suburbanization of America. "J. C. Penney almost certainly knew every customer who came through the doors of his store in Kemmerer, Wyoming," says Michael Ponder. "But that became impractical as things got bigger. Today, though, it's all possible again."

Moreover, says Ponder, the Web makes possible a customer experience that simply cannot be duplicated in the brick-and-mortar world. But not many sites have caught on to this notion, he says. Most still try to reproduce their current model for store design or customer service on a site. It's a common belief, he says, that if you track a customer's visits to an Internet site, you become acquainted with his or her habits and preferences and can begin to make recommendations. But that's rarely effective, Ponder says. If a person makes five purchases, and two of them are gifts for others, what good is the data? He suggests there is a much simpler and more powerful way to figure out how to predict what a customer might want to buy. "Why don't we just ask them?" he wonders. "With the Internet, you can build a site that conforms to the individual no matter how far out of the norm they are."

Much of what Ponder hopes to see remains in the development stage as site developers figure out how to accommodate a customer who might decide, as Ponder notes, to ask for something "whimsical." But there are interim steps. *Yahoo.com,* for instance, allows visitors to customize their portal depending on their use preferences. It's entirely feasible to design almost any site to accomplish that same thing. If you're selling clothing online, for instance, why take up a male customer's bandwidth with banners and specials aimed at clothing? Let him click a preference button that says "show men's clothing only when I return." Rudimentary, but a step toward allowing the customer, not the site proprietor, to shape the experience.

Much of Internet development is focused on making the site as "hands off," from the site owner's perspective, as possible. But technology that will allow a company to step in when appropriate is becoming available. If a customer is pondering a high-ticket purchase, for instance, it might be useful to provide a button that is the electronic equivalent of the bell on the counter. A store agent could quickly answer that call, access a screen that details the customer's path to that page, and then step in via voice link directly over the computer to talk with the customer.

Targeted e-mail

E-mail that is carefully targeted at consumers who have asked to receive it can transcend advertising: It becomes a service. And as the outrage over spam seems to be on the wane as ISPs such as America Online find ways to stop it, targeted e-mail is becoming more widely accepted. Computer seller Quantex, for instance, asks buyers of its PCs and laptops to sign up for a newsletter that offers technical tips, special prices on upgrades, and other news. Other e-commerce sites, such as *Cdnow.com,* send out e-mails that contain reviews of new releases, special discounts, and music picks that reflect the recipient's recent purchases. Each mailing is even headlined with "The CDNOW Update—Douglas Gantenbein's Edition."

Of course, unsolicited e-mail deserves the special place in hell where its purveyors will be assigned ("Hi. I'm Satan. This is your PC. Your first job will be to read through and report on 246,871 pieces of spam. When you're finished, you may use the restroom."). Sending any e-mail message that the recipient did not *specifically* ask for is risky. It's good practice to start each message with the reminder that the recipient asked to receive the notice when he or she visited the site recently. And, even though it may be against your business judgment to do so, make it as simple as possible for the recipient to ask to be taken off the mailing list. Put information on how to do so, in fact, in the first or second paragraph of the message. Few retailers who send such targeted e-mail do so, but this practice could earn you considerable goodwill among those who frequent your site.

A Tale of Two Sites

What does a top-flight Web site look like? It's not the result of flashy graphics. Sure, looks count, but less than you'd think. Visitors want a site that is easy to navigate, fast to load, and knows who they are.

REI: A Community of Outdoor Buffs

Recreational Equipment, Inc., based in Seattle, is one of the most unusual companies in the United States. Founded as a cooperative among friends in the 1930s as a source for then hard-to-find European climbing equipment, REI has grown to more than 1.6 million members, 55 stores, and $590 million in sales. The company remains a cooperative—the largest of its kind in the United States—and each year members receive checks equivalent to about 10 percent of their past year's purchases.

REI's stores reflect its commitment to "muscle-powered" sports and its roots as a climbers' resource. And its Web site (*www.rei.com*) reflects that heritage. It is a hugely busy site, with more than 1 million visits a month. It is straightforward in design and loaded with information, and it has been heavily influenced by input from REI members who visit the site. "We've gotten thousands of e-mails about the site," says Matt Hyde. "From those we've been able to figure out the main interests people have and prioritize what the site does."

As a result, the REI site works to be both familiar and distinctive. It adheres to increasingly accepted Internet standards by placing its navigation bar down the left side of the page and store banner across the top. "We don't want people to relearn the site," says Hyde. To further that goal, Hyde carries out extensive usability testing. And the site is built to the lowest common Internet denominator. Although cable modems and digital subscriber lines are becoming increasingly common, the vast majority of Web users will be chugging along at not much more than 28.8 bits per second for years to come. Extensive white space and relatively few graphics on the home page, for instance, speed download times.

The REI site won't soon win any design awards. In fact, it's pretty cluttered-looking. But by placing a great deal of material on the home page, its designers have made the site "shallow"—a good thing, as it means visitors are typically three clicks away from a product they might be seeking. That's no mean feat, as REI's Internet catalog contains more than 56,000 items, more even than the company's mammoth Seattle flagship store. If a customer is looking for a Marmot Nutshell tent, for instance, he or she simply clicks "camp/hike" on the home page, then "backpacking tents" from a list of broad camping categories, and then the Marmot tent desired. Searchers also can use the site easily, although search filters are employed to ensure that a search doesn't overwhelm the questioner.

Rei.com also works to tap into one of the company's strongest assets: the common interests of its members. Site visitors can access far-ranging tutorials on buying and using outdoor equipment. Climbers, for instance, can visit climbing newsgroups. And shoppers with particular interests can have notifications sent to them when particular kinds of products go on sale or when new products arrive.

eTAC: A Total Customer Support Site

REI has proven itself in the e-commerce arena by offering a broad product range, developing a coherent and easy-to-use site, and leveraging its already loyal customer base. Motorola has taken a very different approach with its new eTAC site (shorthand for Electronic Technical Assistance Center).

Located at *www.eTAC.motorola.com*, the site replaces a half-dozen sites previously used by Motorola's networking and modem customers, most of whom are system administrators and IT professionals. The site offers FAQs, product and ticket status, case history information, and even the ability to purchase technical support services.

Its strength lies in creating a personal experience for each visitor. Site users are assigned a password, and from the very start Motorola captures information that allows the site to automatically tailor itself depending on the visitor. "A principal at a VAR [value-added reseller] has different needs than a technician with the same VAR," says Bob Stewart, program manager for strategic service marketing with Motorola. "Rather than making each person drill down to find the information they what, by knowing who they are we give them a more relevant experience right up front."

Beyond that, the eTAC site strives to let a visitor specify what he or she wants, rather than "pushing" content at a visitor based on what the site operators think a particular user might want. The site has numerous pages that cover such issues as software upgrades, hardware purchases, and technical questions. As a visitor uses the site, he or she can create a custom welcome page that contains the most-used portions of the site.

As with the REI site, eTAC won't win any design awards. But it is clean and easy to understand, follows accepted Internet practice for button placement, and is designed for speed and ease of use more than eye appeal.

Principles of Good Site Design

What makes a Web site that's customer friendly? Volumes could be written on that subject, but Brian Ablaza, a Web designer with The Star Group, a Philadelphia-based advertising agency that also does Web work, offers the following guidelines:

- Offer information—lots of it. "When people come to a site, I'd better have the information they're looking for," he says. "Or they'll just go somewhere else." That's particularly true of e-commerce sites, he says; once people have spent money at your site, their expectations rise.

- Include alternative contact routes. Don't use a Web site as a shield to prevent people from calling or writing to you. "I'd have that contact information right up front on the site," says Ablaza.

- Use generous amounts of white space. It's eye-catching, shows the graphics and headlines you use to maximum advantage, and doesn't soak up bandwidth.

- Do your homework about visitors. Once a site is running, track how people use it and what seems to draw them. "Eighty percent of the people who come to a site need only 20 percent of the information you might have there," Ablaza says. "Use your site logs to figure out where people are going and what they're looking for, then get that information at the front of the site."

- Design a site that is wide and shallow, not narrow and deep. Put lots of navigation buttons on the home page; most people would rather spend a little more time there and then route themselves properly than have to drill through multiple layers to get where they want to go. For instance, rather than having eight navigation buttons on page one, each with eight additional links, Ablaza suggests a "16 by 16" rule—16 navigation buttons with 16 links on each button.

FrontPage 2000 Helps You Become a Webmaster

Microsoft FrontPage 2000 simplifies Web site creation and management for both novice users and Web professionals. It lets users create exactly the site they want by using features such as customizable themes and HTML preservation. FrontPage 2000, for instance, has more than 60 predesigned themes that enable even a novice Web artist to create a useful and visually consistent theme that adheres to standard Web protocols. FrontPage's powerful Reports View function also helps site managers keep a site running smoothly by reporting slow pages, added or changed files, broken hyperlinks, and component errors.

These features, along with a new integrated HTML editor and a seamless integration with Microsoft Office, make Web site creation and management easier than ever.

New features in FrontPage 2000 include the following:

- **HTML preservation.** You can edit existing HTML and scripts with complete source preservation of tag and comment order, capitalization, and even white space.

- **Latest Web technologies.** You can apply cross-browser Dynamic HTML and CSS 2.0 positioning to page elements, restrict the use of specific Web technologies to target browsers, or even add advanced scripts such as Active Server Pages to Web pages.

- **Custom themes.** FrontPage 2000 introduces 10 new business-ready themes among more than 60 professionally designed themes. You can easily create your own themes or customize an existing theme to create a unique look for your Web site.

- **Easy-to-use databases.** Form results can be automatically saved to an Access database, and interactive database queries can be easily incorporated directly into a Web page with just a few mouse clicks.

Put simply, the Internet could be the greatest customer-service innovation since "satisfaction guaranteed." It can allow many customers to literally solve their own problems, freeing up a company's staff to focus on the trickiest and most satisfaction-sensitive issues.

Part 3

Appendixes

Direct-Marketing Resources

Books

The following books are handy reference guides for the novice or advanced direct marketer.

Bacon, Mark S. *Do-It-Yourself Direct Marketing: Secrets for Small Business*. New York, NY, John Wiley & Sons, 1997.

A guide to all aspects of direct marketing that keeps the small-business person in mind.

Baier Stein, Donna, and Floyd Kemske. *Write On Target: The Direct Marketer's Copywriting Handbook*. Lincolnwood, IL, NTC Publishing Group, 1991.

Baier Stein, a well-known direct-marketing expert, and Floyd Kemske, a business writer, focus on the task of writing effective copy.

Bly, Robert. *Power-Packed Direct Mail*. New York, NY, Henry Holt & Co., 1995.

Bly is a veteran marketer who offers valuable advice on direct mail.

Geller, Lois K. *Response! The Complete Guide to Profitable Direct Marketing*. New York, NY, Simon & Schuster, 1996.

Geller, a near-legend in direct-marketing circles, has written an approachable and authoritative guide to direct marketing.

Periodicals

Target Marketing
North American Publishing Co.
Box 401
North Broad Street
Philadelphia, PA 19108
(215) 238-5300
www.targetonline.com

Monthly magazine aimed at users and producers of direct marketing. Focuses on the creation, printing, and mailing of direct mail.

Direct: The Magazine of Direct
Marketing Management
Cowles Business Media
470 Park Avenue South
New York, NY 10016
(212) 683-3540
www.mediacentral.com

Published 16 times a year, Direct *reports on news and trends in the direct-marketing industry. Worth the price of subscription if only for Herschell Gordon Lewis's wonderful columns.*

Direct Marketing Magazine
Hoke Communications, Inc.
224 Seventh Street
Garden City, NY 11530
(516) 746-6700

Useful monthly magazine packed with how-to articles written by experienced direct-mail practitioners.

DM News
Mill Hollow Corp.
100 Avenue of the Americas, 6th Floor
New York, NY 10013
(212) 925-7300
www.dmnews.com

Weekly newspaper of record for the direct-marketing industry. Features news, trends, and how-to articles.

Directories

Standard Rate and Data Service (SRDS)
1700 Higgins Road
Des Plaines, IL 60018
(800) 851-7737
www.srds.com

SRDS publishes the industry standard in direct-mail references. Among them are Direct Marketing List Source, Consumer Magazine Advertising Source, *and* Newspaper Advertising Source.

Direct Marketing Market Place
Reed Elsevier, Inc.
121 Chanlon Road
New Providence, NJ 07974
(800) 323-6772

> *A listing of service firms and suppliers as well as direct-marketing agencies, events, awards, and contests. Cost is $239 for the annual edition.*

National Directory of Mailing Lists

Oxbridge Communications, Inc.
150 Fifth Avenue
New York, NY 10011
www.mediahq.com

> *A guide to 16,000 mailing lists, available in book or CD-ROM form. $545 for the print version; $645 for the CD-ROM.*

Postal Service Guides

To obtain one of these guides, contact your local post office. The well-designed U.S. Postal Service Web site is *www.usps.gov*.

> *Preparing Standard Mail (A)*
> Detailed reference guide on preparing your mail.

> *Quick Service Guide: Mailing Made Easy*
> Detailed explanation of all aspects of mailing.

> *The Small Business Guide to Advertising with Direct Mail*
> Pocket-sized booklet packed with useful tips on direct mail.

Organizations

Direct-mailing clubs and organizations can provide a wealth of information about direct mailing. Most also offer regular programs and classes on all aspects of direct-mail marketing. Clubs are located in most major metropolitan areas in the United States.

Direct Marketing Association
6 East 43rd Street
New York, NY 10017
(212) 768-7277
www.the-dma.org

Regional direct-marketing groups include the following:

Chicago Association of Direct Marketing
(303) 914-8407

Direct Marketing Club of New York
www.dmcny.org

Direct Marketing Club of Southern California
www.dmcsc.com

Florida Direct Marketing Association
www.fdma.org

Midwest Direct Marketing Association
www.mdma.com

New England Direct Marketing Association
www.nedma.com

Rocky Mountain Direct Marketing Association
www.rmdma.com

Seattle Direct Marketing Association
www.sdma.com

Customer-Service Resources

Books

Barlow, Janelle, and Claus Møller. *A Complaint Is a Gift: Using Customer Feedback as a Strategic Tool.* San Francisco, CA, Berrett-Koehler, 1996.

A detailed analysis of how to collect customer complaints and use them to build customer loyalty.

Goltz, Jay. *The Street-Smart Entrepreneur: 133 Tough Lessons I Learned the Hard Way.* Omaha, NE, Addicus Books, 1998.

Refreshingly candid and straightforward information from a guy who built a business from the ground up.

Griffin, Jill. *Customer Loyalty: How to Earn It; How to Keep It.* San Francisco, CA, Jossey-Bass, 1997.

> *A useful seven-step process for making yourself indispensable to customers.*

Karr, Ron, and Don Bohowiak. *The Complete Idiot's Guide to Great Customer Service.* Indianapolis, IN, Macmillan General Reference, 1997.

> *Okay, the name, as always, is off-putting. But this is an easy-to-use reference on customer service.*

O'Dell, Susan M., and Joan Pajunen. *The Butterfly Customer: Capturing the Loyalty of Today's Elusive Consumer.* New York, NY, John Wiley & Sons, 1997.

> *Interesting psychological treatment of what drives the modern customer.*

Whiteley, Richard. *The Customer-Driven Company.* Boston, MA, The Forum Corporation, 1991.

> *A somewhat dated but still authoritative look at how companies succeed by focusing on their customers.*

Periodical

Customer Support Management Magazine
535 Connecticut Avenue
Norwalk, CT 06854
(203) 857-5656
Fax: (203) 899-7417
www.customersupportmgmt.com

Directory

Bacon's Magazine Directory
Annual guide to every consumer and trade publication printed in the United States. Excellent resource for those who wish to learn about an industry by subscribing to trade magazines. Cost is $285 for the combined magazine and newspaper directories. Available from Bacon's Information, Inc., by calling (800) 621-0561 or visiting www.baconsinfo.com.

Organizations

International Customer Service Association (ICSA)
401 North Michigan Avenue
Chicago, IL 60611
(800) 360-4272
www.icsa.com

International organization with some 3500 members that offers a range of resources, including a speakers' bureau, surveys, and member studies.

American Society for Quality (ASQ)
611 East Wisconsin Avenue
P.O. Box 3005
Milwaukee, WI 53201-3005
(800) 248-1946
www.asq.org

The American Society for Quality advances individual and organizational performance excellence worldwide by providing opportunities for learning, quality improvement, and knowledge exchange.

Society of Consumer Affairs Professionals in Business (SOCAP)
801 North Fairfax Street, Suite 404
Alexandria, VA 22314
(703) 519-3700
www.socap.org

The Society of Consumer Affairs Professionals in Business is open to all professionals who are in some way responsible for creating and maintaining customer loyalty.

Index

Smart Business Solutions

Direct Marketing and
Customer Management

DOUGLAS GANTENBEIN

Want to learn about small-business financing? Read on to see a sample chapter from SMART BUSINESS SOLUTIONS FOR FINANCIAL MANAGEMENT, a companion title from Microsoft Press!

If you find SMART BUSINESS SOLUTIONS FOR DIRECT MARKETING AND CUSTOMER MANAGEMENT useful, we'd like to suggest another title that can help you harness the power of technology for your business. The next chapter, excerpted directly from SMART BUSINESS SOLUTIONS FOR FINANCIAL MANAGEMENT (ISBN 0-7356-0682-X), demonstrates how you can get the practical knowledge and skills you need to manage small-business finances efficiently. This title provides complete information on best practices in financial management to help your small business grow and prosper. It also shows how to use the latest technology to your advantage with detailed, real-world examples of accessible financial systems you can quickly set up yourself using popular software. It's the one book about finance every small-business owner and home-office entrepreneur needs to keep a business growing.

SMART BUSINESS SOLUTIONS FOR FINANCIAL MANAGEMENT covers topics such as:

- Setting up an efficient electronic bookkeeping system for a small business with Microsoft Money

- Conducting online banking transactions, such as payroll direct deposits, with Money

- Forecasting and planning revenue with Microsoft Financial Manager

- Preparing financial records to facilitate paying taxes with Financial Manager and Money

- Building an e-commerce system to track Web storefront transactions with Money

Rivka Tadjer, author of SMART BUSINESS SOLUTIONS FOR FINANCIAL MANAGEMENT, has written numerous columns about finance, entrepreneurship, and Web commerce for publications such as *The Wall Street Journal Interactive, The Washington Post, Business Week Online, Barron's Online, Home Office Computing, Small Business Computing,* and *PC Week.* She also appears regularly on Cablevision's "Metro Learning" program in the New York metropolitan area to discuss shopping on the Web and other technology issues. She resides in New York City.

Contents

Part 5

Tricks for Today, Plans for Tomorrow

The Lay of the Land: What You Need to Survive

Chapter 1

What Small Businesses Are Up Against Today

Small businesses and home office dwellers have more to contend with than ever before when it comes to handling finances. Keeping the books and managing investments are just the beginning. The technological age has introduced many new variables: online bill paying, online tax filing, and remote employees and subcontractors to track; Web storefronts to manage; and cash-alternative payment methods (for the payments small businesses accept) to handle. All these new variables, of course, offer enormous benefits, but that doesn't detract from the challenge they pose to entrepreneurs who are already juggling far too much and wearing many hats each day. Still, do keep in mind that although the learning curve for managing finances can be steep, the result is efficiency, which eventually means you get to devote less time to thinking about it—a boon in and of itself.

The learning curve for any or all of these new financial factors tends to go something like this: It starts by sounding overwhelming and expensive, so you put it off. Then, with a little bit of knowledge and insight into the benefits of facing your financial demands, it starts to sound appealing, though daunting. Then you start the process, and it seems like it's all possible. Then you get into the thick of learning how to use software and setting up systems, and it sounds daunting and unattainable again. Then the first month goes by when you don't have to sit down and write a bunch of checks—your automated bill-paying system kicks in—and you're motivated again. Then you see the light at the end of the tunnel—that each system will take patience to set up, but it will be well worth it. Next, the brass ring: You see how methodical the whole process is, and you dig in. Finally, you have an efficient system for each and every financial function in your business.

So, how do we get you from daunted to efficient?

Since it tends to hold true that context sheds light, and having light shed on a subject in turn helps you think clearly, let's start by talking about these New-Age challenges that will affect the way you think about the strategies and systems you want in place. The crux of giving context to financial management in this chapter is really to help you think about what is viable in terms of your budget and your human resources, taking all your factors into consideration before deciding what systems you want to start setting in place. It may well be that you need to start slowly, applying only a few systems detailed in this book and saving others—such as Web storefronts and credit-card acceptance—for later in the year or next year, when your budget allows.

The point is to get everything, including a realistic look at what everything will cost you, out in the open. As you may remember from the Introduction, a crucial factor in financial management is to minimize the kinds of financial surprises that leave you short on cash, or worse, in debt.

So let's demystify the key issues before we get to the nitty-gritty of setting up systems.

Cut to the Chase: What Uncle Sam Wants From You

The first step in understanding the challenges to your finances today is to understand what the IRS thinks about you. The IRS is all about definitions (which we get very specific about in Chapter 5).

The main thing you need to know now is that IRS officials, namely tax auditors, are skeptical when looking at small business and individual returns that include home office deductions and other entrepreneurial expenses. In short, a small business tax return raises red flags at the IRS. The reason is that apparently there is a great deal of fraud, especially by people who claim to work at home and deduct everything they do in life, including vacations and other non-deductibles, in the name of business expenses. (Note: If you own your own home, you may not necessarily want to take the home office deduction every year, because of the tax implications should you sell your home.) Furthermore, taxpayers—both the home office dwellers and small businesses with leased commercial space and employees—overestimate deductions. Even though some of these mistakes are made honestly, they encourage scrutiny from the IRS.

Toeing the Line

Say you live in a city in a small apartment that doubles as your home office. It is very likely that a large portion of your apartment is truly devoted to office space. Yet, if the IRS sees that you're trying to deduct a high percentage of square footage as a home office, they may flag your return for scrutiny or trigger an audit. One year, I tried to get 80 percent of my apartment as a home office deduction past my accountant. My former apartment in New York was small, and 80 percent was an honest, if not conservative, estimate of how much of the space I used for a home office. (We're talking about pretty much every room except my bedroom, bathroom, and kitchen!) Yet my accountant shook his head and asked me to lower that estimation.

At first I was angry, to be honest. I believe very strongly that our society should encourage small business owners and entrepreneurs of any kind. I was being totally truthful about my deduction and struggling to make my entrepreneurial life work while saving money by not renting a separate office, effectively forcing me to live amid a sea of computers and papers in my apartment.

After all, if it weren't for entrepreneurs, the nature of capitalism and competition in this country would change drastically—perhaps even making corporate oligopolies the norm, with very little room for true competition, which, in turn, would result in fewer choices for consumers and less creativity and invention in our society overall. If you look at the technology industry alone, you see how much invention comes from small businesses.

Anyway, my accountant nodded his head in sympathy at my philosophical diatribe about the importance of entrepreneurs in a free-market society, then told me once again to change the percentage on my return. Well, I didn't, but I understood that I was taking a chance. The reason I took the chance—aside from strong conviction—is that my apartment space fit the IRS requirements to the letter.

The point of this anecdote about home office deductions is that there are a million details involved in dealing with the IRS as an entrepreneur. The more careful and detailed you are, the better off you will be. Keep this in mind when you decide what kinds of ledgers you want to set up and what line items you will track.

Pay Taxes Online If You Want the IRS to Love You

The other way of avoiding intense IRS scrutiny is to make sure your tax returns are error-free. Many audits start because an auditor finds an honest mistake. Once the auditor is focused on your return, the IRS might go through it with a fine-toothed comb.

The good news is that the likelihood of an audit is actually less if you file your taxes electronically (which you will learn how to do in Chapter 6). According to H&R Block, which is responsible for handling more than 50 percent of all electronic returns filed with the IRS, the error rate on electronic returns is less than 1 percent, but it is 22 percent on print returns. The reason the error rate is lower for electronic filers is simple: If your return is in electronic form, it means you've keyed all numbers into an electronic form, then let the computer program do the calculations automatically. Checking and double checking are built into the process. Plus, if you've put a deduction in the wrong schedule, the computer program alerts you.

No great cost is associated with either electronic bookkeeping or electronic tax filing, so at the very least—if you do nothing else this book suggests—make these two projects your top priorities for financial management.

One other piece of information to bear in mind if you're starting to worry about scrutiny of your tax returns: You do not have to feel guilty or second-guess yourself if you are audited. As long as you have receipts proving your expenses and have taken legitimate deductions, you'll be fine. What an audit does mean is a headache—and that brings us back to our point about keeping fastidious books electronically and filing electronically to avoid errors.

New-Age Employees and Other Disasters

After getting your mind around the basics of electronic bookkeeping and tax filing, start thinking about some of the other issues small businesses face today. At the top of this list are employees who want lots of freedom (just as you do) and the new people the age of technology will inevitably force you to hire (even if you're a home office dweller).

These two factors are exactly what they sound like: cost centers. So, as you get a feel for the realities of these factors, keep your budget in mind. Some of these cost center items may have to wait.

Give Your People Freedom—Or Lose Them

Let's start with employees who want more freedom. As an entrepreneur, you have a good sense of how the American workforce is changing. More and more, people want to work from their homes, whether they own their own businesses or work for corporations as telecommuters.

Some analysts predict that in the next 10 years, close to 40 percent of the service industry workforce will telecommute. For small business owners with employees, this means you will need to accommodate changing lifestyle needs in order to acquire and maintain good talent. For small businesses in particular, no asset is more valuable than intellectual property—the people.

What are these changing lifestyle needs? Telecommuting is a big one. You can count on the fact that your employees will want to work at least part of the week at home. Whether the reason for wanting to work at home is as simple as easing a long commute or as complicated as having to juggle child care obligations, the result is the same. You will need to provide means for your employees to work at home.

One of the consequences of our telecommunications age is that the workforce at large has "seen the light"; everyone understands very clearly that standards of living can be raised with the aid of technology. The means you will need to provide could include new laptop computers or telecommunications access to your databases so that your employees can get at the reports and other documents that "live" in your office computers. They'll need e-mail and maybe a second phone line for faxing or Internet access. These needs add up to capital expenditures—which you'll need to budget for and then track.

The Right Person Might Not Be in the Right Place

This concept of employee telecommuting goes a step further. Say you need to hire someone with particular skills—maybe a salesperson with a lot of experience in your industry. Say the person you really want isn't local and isn't willing to move, but he or she would be extremely valuable to your business. You could end up with a remote employee, one whose relationship to you is primarily conducted via telecommunications methods, namely e-mail and phone. In this scenario, there could be even greater benefits to you than simply having the right person doing the job. From a financial perspective, you might be able to cut a deal with this potential hire to work as a subcontractor, which of course means you don't need to pay benefits. If you and the subcontractor work out such an agreement, you'll need a system to track subcontractors separately from your regular employees.

When it comes to telecommuting employees, remote employees, or the employees who work at your office every day, even more issues arise: You'll need to pay these folks, and most of them will want direct deposit of their paychecks to their bank accounts. This means setting up accounts in Office 2000's Financial Manager.

You'll start to see that the linchpin of all the financial systems you'll invent for yourself—including the one we'll discuss next, the Web storefront—is Office 2000.

The Web Storefront: Albatross or Money Machine? It's Up to You

Another cost center to plan for is your Web storefront. A Web storefront is a Web site for your business that allows people to buy goods and services from you directly over the Internet. Like any good Web site, your Web storefront is a brochure for your business as well. What distinguishes a Web site that's simply a brochure from a Web storefront is the capability to buy products or services.

Selling goods and services through your Web storefront usually means setting up a credit-card transaction-processing system on your Web site. Say you're a small retailer that sells crafts. Potential craft buyers can surf to your Web site, look at pictures and descriptions of the items you sell, choose the ones they want to buy by "clicking" on them, and then key in credit-card or debit-card numbers (totally securely, of course), which are sent to you over the Iternet with the orders. You fulfill the orders and probably send the customers back e-mail receipts.

You can have a Web storefront of sorts without accepting credit cards. Instead, you can simply take orders for goods over the Internet, then send the goods C.O.D. or have a salesperson call the customer to get the credit-card information. Although you can argue that this is indeed a Web store, it isn't really conducting electronic commerce. It actually makes the whole traditional phone-ordering or fax-ordering concept more complicated by adding another step. You might find this concept worthwhile nonetheless, because a Web site itself is a terrific marketing tool; it gives your business worldwide exposure to people you may never otherwise be able to reach.

However, taking the whole order, from start to finish, over the Internet is often the best way to go. It requires much less in terms of human resources, for instance. One of the greatest assets of Internet sales for small businesses is that they require you to have fewer people manning the phones for orders. Web storefronts also provide new revenue streams. (We go into detail about Web storefronts and how to set them up in Chapters 12 through 17.)

What Will It Cost?

The thing you need to remember about Web storefronts is that they cost money and require effort to set up and run. The good news is that over time, they've gotten much cheaper to implement. Here's the bottom line: You can get a Web storefront up and running for about $300 per month. A Web-hosting company (which may or may not be an Internet service provider) will host the whole thing for you, plus provide the Internet access you'll need. That access rate will need to be faster than simply dialing in on a regular phone line. You will need a digital line of some kind, as explained in the Web storefront part of this book. A Web-hosting company will also arrange for your back-end accounting system, which routes the credit-card purchases directly from your Web storefront to the banking system. These same Web-hosting companies often provide reporting as part of the service. This means the host will send you statements of all sales made to integrate into Financial Manager. The reports also tell you valuable marketing information, such as what items are selling best and who's buying what.

The total monthly cost for a Web store varies according to your Internet access speed. Just keep in mind that you'll be spending $300 per month (including your Internet access), at least, when all is said and done.

Will It Pay Off?

How to gauge whether this expenditure will pay off for your business is the key issue for Web storefronts. Knowing the estimated return on your investment will play a part in how you negotiate with Web-hosting services. For example, most Web-hosting services will offer you a better deal per month if you sign on for a full year. However, it may be smarter for you to pay a slightly higher per-month fee in order to test the waters for three months or so to see if you sell anything at all.

Getting a headache yet from all this new information? Don't fret, and don't try to assimilate all these bits of information in your head right now. All the factors related to Web storefronts will be made clear and given to you in bite-sized, easily digestible morsels in the Web storefront section of this book. By the time you're done reading that material, you'll have a step-by-step guide of what to do and when to do it for maximum profit and minimal expense to your business.

The only thing you should make a mental note of right now is that $300 monthly cost. It's a realistic cost. Toss out the lowball figures, such as less than $100 per month, that you've heard at cocktail parties. That kind of unrealistic price leaves some vital component out of the equation, such as Internet access costs or the price you pay as a merchant that accepts credit cards over the Internet. Any credit-card payments you accept cost you. The catch with Web storefronts is that the fee the credit-card companies charge for Internet sales is higher than the fee they charge in the physical world. So, say Visa charges 1.5 percent per sale for items consumers buy in a physical-world retail store. The percentage may be 2.5 percent for similar items sold via the Internet.

Visa, MasterCard, and American Express officials attempt to explain this discrepancy by saying that their risks are different on the Internet because it is a less proven, and therefore higher-risk, sales vehicle. This reasoning is what I affectionately refer to as "banking industry logic"—a breed of reasoning formed by what I believe are extraterrestrial beings who have somehow earned MBAs. Like it or not, though, it is a fact of life. These same credit-card companies offer "reassurance" by claiming that as the Internet proves itself as a marketplace and they gather more data showing payment patterns on which they can base predictions (as they do in traditional retail sales), they might lower that percentage to be on par with physical-world sales.

Why credit-card companies can't use the predictions they make about over-the-phone credit-card sales and apply them to the Internet is beyond me. Essentially, the Internet is a phone network, with Web sites as text-based (instead of voice-based) message systems. Your computer is the phone. This analogy makes sense to the telecommunications industry, but alas, bankers are a different breed.

My point is that there are costs people leave out when they give you too-good-to-be-true estimates for Web storefront costs. You may want to kill the messenger—which would be me—for giving you less optimistic news about per-month fees, but at least you won't have nasty surprises later. My estimates will be your actual costs; I won't leave anything out, so you won't experience sticker shock once your Web-hosting company sends you the real, live bill.

If it ends up costing you a little less than my prediction, well, that's the kind of financial surprise that tends to sit well.

While we're on the subject of monthly costs for new things, let's go to the last cost center: new employees you're going to have to hire.

Technology Means You Need to Hire People, Like It or Not

No matter how much of an electronic do-it-yourselfer you become after reading this book, I cannot save you from one inevitable cost center: technical support staff. One truism about technology is that things inevitably go wrong. Another truism is that small businesses can't do everything themselves—we just got through talking about hiring a Web-hosting company, for instance.

The Catch-22 is that even if you arranged to implement a Web storefront in-house, you'd absolutely, positively—no question about it!—need an in-house technical person on staff to make sure it runs correctly. Don't listen when vendors tell you their systems are so easy to use that you won't need a technical support person. They're wrong.

And that's the point here: These technological benefits mean you will need to hire support staff to handle technical tasks—whether it's contracting a Web-hosting company, hiring a part-timer for tech support, or simply paying extra (up to $1,000 per year at the low end) for access to emergency on-site computer repair services.

Technical support staff doesn't become a need solely if you have a Web storefront (though that makes support staff particularly imperative). The minute you rely on computer systems to run your business—even if it's just loading Office 2000 on your system and keeping your books on it—you need to be able to keep that system up and running all the time, because if your computer network goes down, so does your business. Home office dwellers with one or two computers, a printer, and Internet access are no exceptions.

Thinking about computer tech support as something on which you'll spend $1,000 per year is a wise way to plan ahead. You can buy contracts for such support from the people you buy computer systems from. This kind of arrangement is usually called a *maintenance contract*. If you think computer maintenance is not something worth budgeting for, you should consider the possibility (or the likelihood!) of some kind of computer disaster that will grind your business to a halt—a hard drive crash, a software program on which you store everything you need and which suddenly won't

11

open up—at which time you will be flipping through the Yellow Pages, desperately trying to find anyone who can fix your problem. Such a crisis management service, if you find one, will cost you about $95 per hour, plus a fee just for someone to show up at your door.

You will pay less if you think ahead. Maybe it will cost you $1,000 per year, or $2,000, or for some home offices, $500 or so. The price will vary according to the complexity of your computer network setup. But whatever you pay planning ahead will be less than you would pay dealing with computer repair people once you ran into an emergency.

Bottom line: The day you start running your business electronically is the day you add a technical maintenance line item to your budget. Part of this maintenance budget might go for backup systems—Zip drives, for instance. Once you have a technical support person advising you, you'll know what you need for your situation.

The benefits to these expenditures are many. Perhaps you will even be able to upgrade less often because your systems will be better maintained. Simply cleaning a computer system properly can prevent problems such as hard drive crashes.

So, add $500 to $1,000 per year to your budget for computer maintenance. Later chapters go into detail about how to find these mysterious technical support people. You'll have a resource list at your fingertips by the time you're done reading this book.

With This Context in Mind, Time to Move On

Now, without further delay, let's get to the systems and how they work. Let me quickly outline what you'll be looking at in the next few chapters.

The next chapter, Chapter 2, is devoted to telling you about the features of Office 2000 and the Microsoft Money program. The chapter starts with a detailed description of these two programs because they are the linchpins of setting up your bookkeeping system. Once you understand what these programs offer, it will become clear how each financial management function is tracked as you read the various sections of this book. You'll have context needed to understand what you're reading, and you'll know the capabilities of the software.

Chapter 3 takes you step by step through the process of installing the Money software and setting up your accounts. Chapters 3 and 4 discuss issues you'll want to consider when you set up your books. So let's get started.

Optimize
Microsoft Office 2000
with multimedia training!

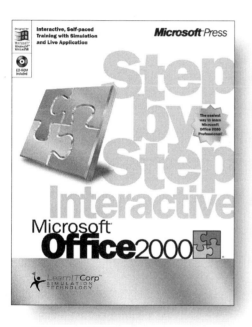

MICROSOFT® OFFICE 2000 STEP BY STEP INTERACTIVE is a multimedia learning system (in both audio and text versions) that shows you, through 20 to 30 hours of simulated live-in-the application training, how to maximize the productivity potential of the Office 2000 programs: Microsoft Excel 2000, Word 2000, Access 2000, PowerPoint® 2000, Outlook® 2000, Publisher 2000, and Small Business Tools. If you already use Microsoft Office 97, this learning solution will help you make the transition to Office 2000 quickly and easily, and reach an even greater level of productivity.

U.S.A.	**$29.99**
U.K.	$27.99 [V.A.T. included]
Canada	$44.99
ISBN 0-7356-0506-8	

Microsoft Press® products are available worldwide wherever quality computer books are sold. For more information, contact your book or computer retailer, software reseller, or local Microsoft Sales Office, or visit our Web site at mspress.microsoft.com. To locate your nearest source for Microsoft Press products, or to order directly, call 1-800-MSPRESS in the U.S. (in Canada, call 1-800-268-2222).

Prices and availability dates are subject to change.

Microsoft®

mspress.microsoft.com

Get technical
help and support—
direct from Microsoft.

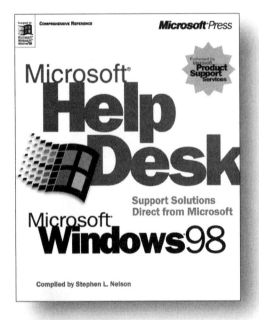

The Microsoft Windows 98 operating system is the upgrade to Windows that makes computers work better and play better. MICROSOFT® HELP DESK FOR MICROSOFT WINDOWS® 98 puts a portable, rich source of Microsoft product support solutions at your fingertips. Written in the clear, understandable language characteristic of the Microsoft Help Desk series, the book covers key Microsoft Windows 98 support issues and their solutions. It is the only Windows 98 Help Desk resource based directly on the archives of Microsoft Product Support's KnowledgeBase.

U.S.A.	**$49.99**
U.K.	£45.99
Canada	$74.99
ISBN 0-7356-0632-3	

mspress.microsoft.com

See clearly—
now!

Here's the remarkable, *visual* way to quickly find answers about the power-fully integrated features of the Microsoft® Office 2000 applications. Microsoft Press AT A GLANCE books let you focus on particular tasks and show you, with clear, numbered steps, the easiest way to get them done right now. Put Office 2000 to work today with AT A GLANCE learning solutions, made by Microsoft.

- MICROSOFT OFFICE 2000 PROFESSIONAL AT A GLANCE
- MICROSOFT WORD 2000 AT A GLANCE
- MICROSOFT EXCEL 2000 AT A GLANCE
- MICROSOFT POWERPOINT® 2000 AT A GLANCE
- MICROSOFT ACCESS 2000 AT A GLANCE
- MICROSOFT FRONTPAGE® 2000 AT A GLANCE
- MICROSOFT PUBLISHER 2000 AT A GLANCE
- MICROSOFT OFFICE 2000 SMALL BUSINESS AT A GLANCE
- MICROSOFT PHOTODRAW™ 2000 AT A GLANCE
- MICROSOFT INTERNET EXPLORER 5 AT A GLANCE
- MICROSOFT OUTLOOK® 2000 AT A GLANCE

Microsoft®

mspress.microsoft.com

Stay in the *running* for maximum productivity.

These are *the* answer books for business users of Microsoft® Office 2000. They are packed with everything from quick, clear instructions for new users to comprehensive answers for power users—the authoritative reference to keep by your computer and use every day. The Running series—learning solutions made by Microsoft.

- RUNNING MICROSOFT EXCEL 2000
- RUNNING MICROSOFT OFFICE 2000 PREMIUM
- RUNNING MICROSOFT OFFICE 2000 PROFESSIONAL
- RUNNING MICROSOFT OFFICE 2000 SMALL BUSINESS
- RUNNING MICROSOFT WORD 2000
- RUNNING MICROSOFT POWERPOINT® 2000
- RUNNING MICROSOFT ACCESS 2000
- RUNNING MICROSOFT INTERNET EXPLORER 5
- RUNNING MICROSOFT FRONTPAGE® 2000
- RUNNING MICROSOFT OUTLOOK® 2000

Microsoft®

mspress.microsoft.com

Register Today!

Return this
*Smart Business Solutions for Direct Marketing
and Customer Management*
registration card today

Microsoft Press

mspress.microsoft.com

OWNER REGISTRATION CARD

0-7356-0683-8

Smart Business Solutions for Direct Marketing and Customer Management

_____ _____ _____
FIRST NAME   MIDDLE INITIAL   LAST NAME

_____
INSTITUTION OR COMPANY NAME

_____
ADDRESS

_____

_____ _____ _____
CITY   STATE   ZIP

_____ ( )_____
E-MAIL ADDRESS   PHONE NUMBER

U.S. and Canada addresses only. Fill in information above and mail postage-free.
Please mail only the bottom half of this page.

For information about Microsoft Press®
products, visit our Web site at
mspress.microsoft.com

Microsoft®*Press*

NO POSTAGE
NECESSARY
IF MAILED
IN THE
UNITED STATES